Organic Chemistry: A Very Short Introduction

VERY SHORT INTRODUCTIONS are for anyone wanting a stimulating and accessible way into a new subject. They are written by experts, and have been translated into more than 45 different languages.

The series began in 1995, and now covers a wide variety of topics in every discipline. The VSI library now contains over 500 volumes—a Very Short Introduction to everything from Psychology and Philosophy of Science to American History and Relativity—and continues to grow in every subject area.

Very Short Introductions available now:

Available soon:

SHAKESPEARE'S TRAGEDIES
  Stanley Wells
CLINICAL PSYCHOLOGY
  Susan Llewelyn and
  Katie Aafjes-van Doorn

EUROPEAN UNION LAW
  Anthony Arnull
BRANDING Robert Jones
OCEANS
  Dorrik Stow

For more information visit our website

www.oup.com/vsi/

Graham Patrick

ORGANIC CHEMISTRY

A Very Short Introduction

OXFORD

UNIVERSITY PRESS

Great Clarendon Street, Oxford, OX2 6DP,
United Kingdom

Oxford University Press is a department of the University of Oxford.
It furthers the University's objective of excellence in research, scholarship,
and education by publishing worldwide. Oxford is a registered trade mark of
Oxford University Press in the UK and in certain other countries

© Graham Patrick 2017

The moral rights of the author have been asserted

First edition published in 2017

Published in the United States of America by Oxford University Press
198 Madison Avenue, New York, NY 10016, United States of America

British Library Cataloguing in Publication Data
Data available

Library of Congress Control Number: 2016955808

ISBN 978-0-19-875977-5

Printed and bound by
CPI Group (UK) Ltd, Croydon, CR0 4YY

Contents

List of illustrations

© 123RF/ efks

Chapter 1
Introduction

Organic chemistry is a branch of chemistry that studies carbon-based compounds in terms of their structure, properties, and synthesis. In contrast, inorganic chemistry covers the chemistry of all the other elements in the periodic table (Figure 1).

This raises the question why one of the three main fields of chemistry is related purely to carbon-based compounds. One answer lies in the fact that carbon-based compounds are crucial to the chemistry of life. Indeed, the term 'organic chemistry' was first mooted in the 18th century by a Swedish chemist called Torbern Bergman to define the chemistry of compounds derived from living organisms. Scientists at that time believed that the chemicals of life (biochemicals) were different from those produced in the laboratory because they contained a special property that only life could provide.

To be fair, there was some justification for this belief. The biochemicals that had been identified at that time had proved difficult to isolate from living systems, and had decomposed more quickly than inorganic chemicals isolated from minerals. Therefore, it was concluded that organic compounds contained a 'vital force' that could only originate from a living organism. Consequently, it was logical to conclude that biochemicals could

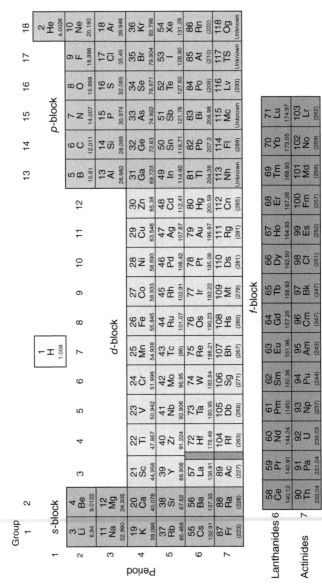

1. The periodic table.

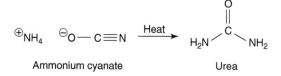

2. Synthesis of urea.

not be synthesized in the laboratory. However, it was not long before this vitalistic theory was challenged. Urea (Figure 2) is a crystalline compound that can be isolated from urine. According to vitalistic theory, it should be unique to life, but, in 1828, it was discovered that it could be synthesized by heating an inorganic salt called ammonium cyanate.

Since then, organic chemistry has come to be defined as the chemistry of carbon-based compounds, whether they originate from a living system or not. Nevertheless, the chemistry of carbon-based compounds is very much associated with the chemistry of life, and the phrase 'carbon-based life forms' reflects that fact. We shall be exploring the importance of organic chemistry to living systems in Chapter 4—a branch of science now defined as biochemistry.

There is another reason for considering organic chemistry as a specialist field, and that is because of the vast numbers of different organic compounds that can be synthesized—far more than would be possible for any other element. Indeed, it has been calculated that the number of different medium-sized organic molecules that could be synthesized amounts to 10^{63}. This is a vast number, so vast that there is not sufficient carbon in the universe to achieve that goal. Note also that these figures are based on medium-sized molecules containing less than thirty carbon atoms. It ignores all the polymers that are possible. In truth, there is virtually an infinite number of novel compounds that could be synthesized, the vast majority of which have never existed on this planet. To date, 16 million compounds have been synthesized in organic chemistry laboratories across the world, with novel compounds being synthesized every day. This still represents a pinprick in the number of structures that could be synthesized. There is plenty of scope for further innovation—a fact that both excites and motivates organic chemists searching for new molecules for new purposes, whether that be in medicine, agriculture, consumer goods, or material science.

For over a hundred years, organic chemists have contributed vastly to our understanding of life at the molecular level, and produced novel compounds that have revolutionized modern society. The fruits of this research can be found in the clothes we wear, the houses we live in, and the food that we eat. The list of commodities that rely on organic chemistry include plastics, synthetic fabrics, perfumes, colourings, sweeteners, synthetic rubbers, and many other items that we use every day. It has produced the insecticides, herbicides, and fungicides that allow farmers to produce sufficient food for an ever-increasing world population, as well as the medicines that tackle disease and prolong lifespans.

The benefits have been enormous, but it is also important to appreciate the drawbacks. New discoveries can produce problems that affect health and the environment if they are not used with due care and responsibility. Unfortunately, such problems can lead to a distrust of new technology in general and chemicals in particular—an attitude that has been defined as chemophobia. The very word 'chemical' is considered by some to indicate a toxic or polluting compound synthesized by the chemical industry. The truth is that the word 'chemical' is a generic term that covers both natural and synthetic compounds. There is also a misinformed belief that synthetic compounds are inherently dangerous, whereas natural chemicals are much safer. Nothing could be further from the truth. Some of the most lethal toxins known to science come from the natural world, while many synthetic compounds are extremely safe. It is also not fully appreciated that a naturally occurring compound synthesized in the laboratory is no different from the same compound extracted from a natural source.

It is true that many of the novel compounds that were introduced for the benefit of society have produced long-term problems, but that does not mean that society should turn its back on all the medicines, pesticides, food additives, and polymers that it relies on. Instead, the challenge is to design better compounds with improved properties. It is the responsibility of organic chemists to learn from any mistakes of the past and to continue striving for the discoveries that will benefit us all. This book illustrates many of the enormous benefits that have resulted from organic chemistry research, as well as some of the problems that have arisen from past innovations. It also shows how today's researchers are seeking a new generation of safer, more effective compounds.

Chapter 2
The fundamentals

Carbon: an elemental socialite

In Chapter 1, it was stated that the number of possible carbon-based compounds is so vast that it can never be achieved. This huge number of structures is given the rather unusual term chemical space. In a sense, there is an analogy between exploring the universe and exploring the synthesis of new organic compounds. In both cases, it represents an endless task, but one which is full of exciting discoveries. In this section, we investigate why the element carbon, rather than any other element, is so suited for the generation of so many different compounds.

Carbon has atomic number 6, which means that it has six protons located in its nucleus. For a neutral carbon atom, there are six electrons occupying the space around the nucleus (Figure 3).

These electrons occupy two different shells (or orbits) around the nucleus. The first inner shell contains two electrons, which is the maximum number of electrons that it can accommodate, while the second shell (the outer shell) takes the remaining four electrons. The electrons in the outer shell are defined as the valence electrons and these determine the chemical properties of the atom. The valence electrons are easily 'accessible' compared to the two electrons in the first shell. The inner-shell electrons

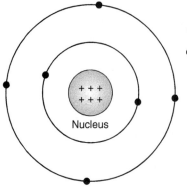

+   represents a proton

●   represents an electron

3. The carbon atom.

are closer to the nucleus, and hidden by the second shell of electrons.

There is great significance in carbon being in the middle of the periodic table. Elements which are close to the left-hand side of the periodic table can lose their valence electrons to form positive ions. For example, lithium can lose its only valence electron to form a positive ion (Li^+), magnesium can lose its two valence electrons to form a magnesium ion (Mg^{2+}), while aluminium can lose its three valence electrons to form an aluminium ion (Al^{3+}). Elements on the right-hand side of the table can gain electrons to form negatively charged ions. For example, fluorine can gain one valence electron to give a fluoride ion (F^-), while oxygen can gain two electrons to form an oxide ion (O^{2-}). The impetus for elements to form ions is the stability that is gained by having a full outer shell of electrons. For example, a fluoride ion has a full complement of eight electrons in its outer shell. Similarly, when a lithium ion loses its single valence shell electron, it is left with a full inner shell of electrons.

Ion formation is feasible for elements situated to the left or the right of the periodic table, but it is less feasible for elements in the middle of the table. For carbon to gain a full outer shell of electrons, it would have to lose or gain four valence electrons, but this would require far too much energy. Therefore, carbon achieves a stable, full outer shell of electrons by another method. It shares electrons with other elements to form bonds. Carbon excels in this and can be considered chemistry's ultimate elemental socialite. Instead of living a solitary existence as an atom or an ion, a carbon atom forms bonds with other atoms to form atomic networks called molecules. The atoms are linked together by covalent bonds, with each bond containing two electrons shared between the two atoms involved.

One of the simplest organic molecules is methane, where a carbon atom shares its four valence electrons with four hydrogen atoms. Similarly, each of the hydrogen atoms shares its single valence electron with the carbon atom. Each bond is made up of two electrons—one from each atom involved in the bond. By sharing valence electrons, each atom in the molecule has a full outer shell of electrons (Figure 4). Unlike an ion, the molecule has no charge.

A covalent bond can also be formed between two carbon atoms. For example, ethane has a covalent bond between its two carbon atoms, as well as six covalent bonds between the carbon and hydrogen atoms (Figure 5).

Carbon's ability to form covalent bonds with other carbon atoms is one of the principle reasons why so many organic molecules are possible. Carbon atoms can be linked together in an almost limitless way to form a mind-blowing variety of carbon skeletons. These include linear chains, branched chains, rings, and combinations of all three. However, the variety does not stop there. Carbon can form covalent bonds to a large range of other elements. We have already seen that carbon can form a bond to hydrogen, but it can also form bonds to atoms such as nitrogen, phosphorus, oxygen, sulphur, fluorine, chlorine, bromine, and

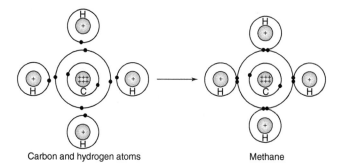

Carbon and hydrogen atoms   Methane

4. Methane.

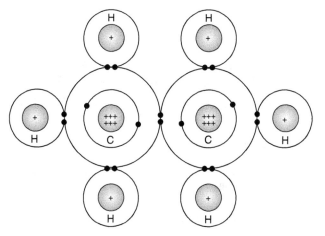

5. Ethane.

iodine. As a result, organic molecules can contain a variety of different elements. Further variety can arise because it is possible for carbon to form double bonds or triple bonds to a variety of other atoms. The most common double bonds are formed between carbon and oxygen, carbon and nitrogen, or between two carbon atoms. One example is methanal (formaldehyde) (Figure 6a). The most common triple bonds are found between carbon and

(a)

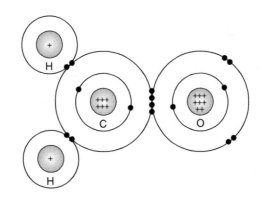

(b)

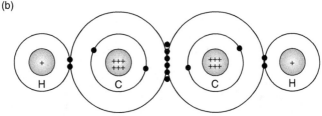

6. (a) Methanal and (b) ethyne.

nitrogen, or between two carbon atoms. An example is ethyne (acetylene) (Figure 6b).

Naming compounds and identifying their structure

Each organic compound is given a specific name that accurately defines its structure using nomenclature rules determined by the International Union of Pure and Applied Chemistry (IUPAC). The more complex the structure, the more complex the name. For example, the IUPAC name for a well-known steroid hormone is (8R,9S,13S,14S,17S)-13-methyl-6,7,8,9,11,12,14,15,16,17-decahydrocyclopenta[a]phenanthrene-3,17-diol. This is quite a mouthful, and so well-known compounds having fairly complex names are often identified by more user-friendly names. For example, the steroid hormone with the long IUPAC name is more commonly known as estradiol. A lot of biologically important compounds are generally identified by their common name; for example morphine, haemoglobin, and adrenaline.

Organic chemists are specifically interested in molecular structure. In the same way that an architect is interested in the structure of a building and uses plans to visualize that building, a chemist is interested in molecular architecture, and how the atoms present are linked together. Therefore, a structural representation of a molecule often means more to a chemist than its name. We have already seen a method of drawing structures in Figures 3–6, but diagrams like this take a long time to draw. A simpler method involves using a line to represent each bond, and the element symbol to represent each atom. For example, structural representations of methane, ethane, and estradiol are shown in Figure 7.

These simplified diagrams identify all the atoms that are present and the way in which they are linked. However, this approach is still unwieldy when trying to represent complex molecules such as estradiol. An even simpler 'shorthand' approach is to omit the carbon atom labels, and to leave out the hydrogen atoms and their bonds. This method cannot be used for methane, since the

Methane

Ethane

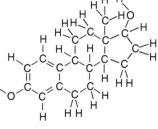

Estradiol

7. **Structures of methane, ethane, and estradiol.**

Ethane

8. 'Shorthand' structural diagrams for ethane and estradiol.

structure would end up being represented by a dot, but it can be used for ethane and estradiol (Figure 8). In structures such as these, a carbon atom is understood to be present at the end of each line, as well as at each corner. The exception to this rule is when an elemental symbol is indicated, as in the two hydroxide (OH) groups present in estradiol. The number of hydrogen atoms attached to each carbon atom can be worked out by appreciating that each carbon atom must have four bonds. If it has fewer than four, the missing bonds are assumed to be to hydrogen atoms. This can be demonstrated by comparing the structures of ethane and estradiol in Figures 7 and 8.

There are several advantages in representing molecules in this way. First, they are much quicker to draw. Second, it is easier to identify the molecular skeleton. As an analogy, the skeleton of a tree is difficult to pick out in the summer because of its leaves, but easily discernible in winter when the leaves have fallen. As far as molecules are concerned, the hydrogen atoms correspond to the leaves. A third advantage of drawing molecules in this way is the ease with which functional groups (discussed later on in this chapter) can be identified.

Stereochemistry

Molecules are three-dimensional objects with particular shapes. Carbon atoms in a molecule can be described as tetrahedral,

Tetrahedron

Tetrahedral shape
of methane

Methane

9. The tetrahedral shape of methane.

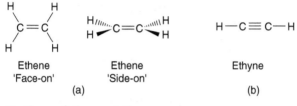

Ethene
'Face-on'

Ethene
'Side-on'

(a)

Ethyne

(b)

10. Shapes of ethene and ethyne.

trigonal, or digonal, but this can be a bit misleading since it is not the carbon atoms themselves that have these shapes. Instead, the shapes refer to the arrangement of the bonds around the carbon atoms. Thus, methane has a central carbon atom with its four bonds pointing to the corners of a tetrahedron (Figure 9). When drawing methane, simple lines indicate the orientation of bonds that are in the plane of the paper. Solid wedge-shaped bonds represent bonds that are pointing out of the page towards the viewer. Hatched wedge-shaped bonds represent bonds that are pointing behind the page away from the viewer. In general, carbon atoms having four single bonds are described as tetrahedral carbons and the bond angles are approximately 109°.

When a carbon atom is part of a double bond, it is described as trigonal, and the bonds around it are in the same plane. For example, both carbon atoms in ethene are trigonal planar, making the overall molecule planar in shape (Figure 10). The bond angles

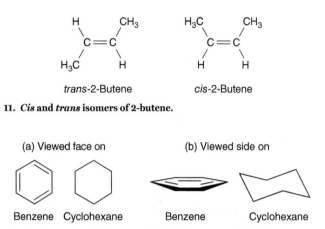

trans-2-Butene   *cis*-2-Butene

11. *Cis* and *trans* isomers of 2-butene.

(a) Viewed face on   (b) Viewed side on

Benzene   Cyclohexane   Benzene   Cyclohexane

12. **Benzene and cyclohexane viewed (a) face on and (b) side on.**

are larger than those of a tetrahedral carbon at 120°. Carbon atoms involved in a triple bond are described as digonal. Here, the bond angle is 180°, and so ethyne (also known as acetylene) is linear in shape.

The double bond in ethene is rigid and cannot rotate. This has important consequences in terms of stereochemistry if there are substituents at each end of the double bond. For example, two different structures are possible for 2-butene (Figure 11). These are known as the *cis* and the *trans* isomers. In the *cis* isomer, the methyl substituents are on the same side of the double bond, whereas in the *trans* isomer, they are on opposite sides. The two isomers cannot interconvert because of the rigid double bond and are different compounds with different chemical and physical properties.

It is possible to get organic molecules where the carbon atoms are linked together to form rings. These, too, can have distinctive shapes. For example, benzene and cyclohexane are both 6-membered rings (Figure 12). The carbon framework for

14

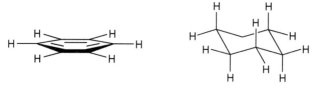

Benzene   Cyclohexane

13. Side-on views of benzene and cyclohexane with hydrogen atoms included.

cyclohexane is made up of six single bonds, whereas the carbon framework for benzene appears to be made up of alternating single and double bonds. In Figure 12a, the two rings look identical in shape, but if they are viewed 'side on', as in Figure 12b, the benzene ring is seen to be planar and the cyclohexane ring is puckered into what is termed a chair shape. This is a consequence of the different bond angles of the carbon atoms involved. The carbon atoms in cyclohexane have bond angles of 109° as in methane and ethane. In benzene, the bond angles are 120° as in ethene.

The difference in shape between the two molecules is further emphasized if the hydrogen atoms are shown (Figure 13). In benzene, the hydrogen atoms are in the same plane as the ring, whereas in cyclohexane they are pointing in different directions. This means that cyclohexane is a much bulkier molecule.

The carbon–carbon bond lengths in cyclohexane are identical, which is to be expected since they are all single carbon–carbon bonds. Curiously, the carbon–carbon bond lengths in benzene are also identical. This would not be expected if the ring was made up of alternating single and double bonds, since double bonds are known to be shorter than single bonds. This is evidence that there is more to benzene than meets the eye. In fact, six of the electrons involved in these double bonds are shared round the ring as a whole—a process known as delocalization. This gives the benzene

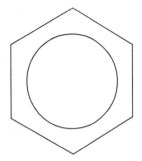

14. Representation of benzene indicating delocalization of electrons.

ring greater stability than a molecule containing three distinct double bonds. The delocalization of the electrons in benzene is sometimes indicated by representing the benzene ring with a circle in the middle to show that the six electrons have mobility round the ring (Figure 14).

The stability of the benzene ring means that it is found in many natural products, and, whenever it is present, it indicates a planar region of the molecule. For example, Figure 15 illustrates a three-dimensional structure of the hormone estradiol from two different perspectives. It is clear that the region of the molecule containing the benzene ring is planar, whereas the rest of the molecule is much bulkier. The planar region has a crucial role in the biological activity of estradiol. Estradiol has to bind to a protein in the body in order to have its hormonal activity. This is only possible because the benzene ring fits into a narrow slot in the protein, something that would be impossible with a bulkier ring.

Wedge-shaped bonds are often used to give an impression of a molecule's 3D shape (Figures 9, 10, 12, and 13). They are also important in defining the relative orientation of bonds where there might be some ambiguity. For example, the structure of estradiol is often represented as shown in Figure 16. The

16

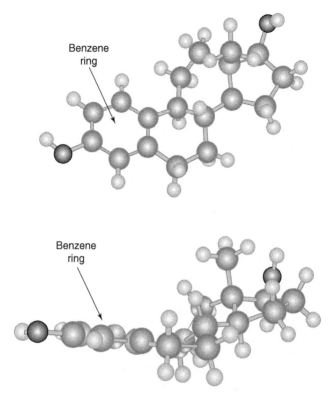

Benzene
ring

Benzene
ring

15. **The shape of estradiol from two different perspectives.**

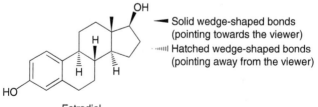

OH

◄── Solid wedge-shaped bonds
(pointing towards the viewer)

⸱⸱⸱⸱⸱⸱⸱ Hatched wedge-shaped bonds
(pointing away from the viewer)

HO

Estradiol

16. **Structure of estradiol including wedge-shaped bonds.**

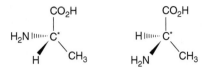

17. The two enantiomers of alanine.

wedge-shaped bonds define the shape (or stereochemistry) at key positions known as chiral centres.

A chiral carbon centre is a tetrahedral carbon atom with four single bonds to four different substituents. For example, the amino acid alanine has a chiral centre indicated by the star in Figure 17.

Any molecule with a chiral centre is defined as being chiral or asymmetric. In other words, it lacks symmetry. There are two possible structures for such molecules, each one being a non-superimposable mirror image of the other. An asymmetric molecule must be one mirror image or the other—it cannot be both. Neither can it switch from one mirror-image structure to the other. These mirror structures are called enantiomers, and the wedge-shaped bonds are essential in distinguishing one from the other. In terms of chemistry, two enantiomers behave identically in their reactions with common chemical reagents. They also have identical physical properties.

At first sight, this might suggest that chirality is of little consequence. In fact, chirality has huge importance. The two enantiomers of a chiral molecule behave differently when they interact with other chiral molecules, and this has important consequences in the chemistry of life. As an analogy, consider your left and right hands. These are asymmetric in shape and are non-superimposable mirror images. Similarly, a pair of gloves are non-superimposable mirror images. A left hand will fit snugly into a left-hand glove, but not into a right-hand glove. In the molecular world, a similar thing occurs. The proteins in our bodies are chiral molecules which can distinguish between the enantiomers of other molecules. For example, enzymes can distinguish between the two enantiomers of a chiral compound and catalyse a reaction with one of the enantiomers but not the other.

Functional groups

A key concept in organic chemistry is the functional group. A functional group is essentially a distinctive arrangement of atoms and bonds. There are hundreds of different types of functional groups, and some of the more common ones are shown in Figure 18.

Functional groups react in particular ways, and so it is possible to predict how a molecule might react based on the functional groups that are present. For example, a carboxylic acid and a phenol both contain an acidic hydrogen that can be lost in the presence of a base to produce a negatively charged ion (Figure 19a and c). An amine functional group is basic in nature and can be protonated to give a positively charged ion (Figure 19b).

These properties are extremely useful in practical organic chemistry as it is possible to separate compounds containing carboxylic acids, phenols, or amines from other types of organic compound in a process known as extraction (see also Figure 35). Compounds containing carboxylic acids or phenols will dissolve in aqueous base, whereas compounds containing amines will dissolve in aqueous acid. Organic compounds containing neither of these functional groups are generally insoluble in water.

By taking advantage of these properties, it is possible to extract compounds containing carboxylic acids, phenols, and amines from synthetic mixtures or from plant extracts.

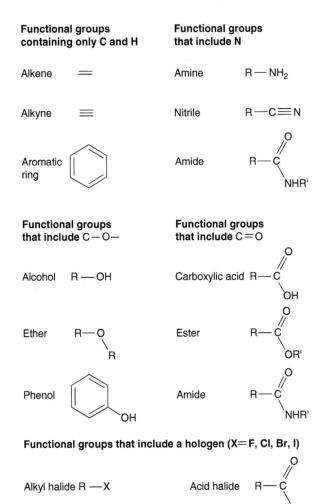

Functional groups containing only C and H

Alkene ═

Alkyne ≡

Aromatic ring

Functional groups that include N

Amine R — NH₂

Nitrile R—C≡N

Amide R—C(=O)NHR'

Functional groups that include C—O—

Alcohol R — OH

Ether R—O—R

Phenol (OH)

Functional groups that include C═O

Carboxylic acid R—C(=O)OH

Ester R—C(=O)OR'

Amide R—C(=O)NHR'

Functional groups that include a hologen (X═F, Cl, Br, I)

Alkyl halide R —X

Acid halide R—C(=O)X

18. **Examples of common functional groups (R indicates the rest of the molecule).**

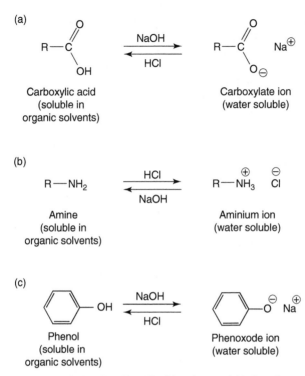

(a)

Carboxylic acid
(soluble in
organic solvents)

Carboxylate ion
(water soluble)

(b)

Amine
(soluble in
organic solvents)

Aminium ion
(water soluble)

(c)

Phenol
(soluble in
organic solvents)

Phenoxode ion
(water soluble)

19. Ionization of (a) carboxylic acids, (b) amines, and (c) phenols.

Intermolecular and intramolecular interactions

The covalent bonds that link atoms *within* a molecule are strong
and are not easily broken. However, weaker forms of bonds can
exist *between* molecules. These are defined as intermolecular
bonds. The main ones are hydrogen bonding, London dispersion
forces (also known as van der Waals interactions), and ionic
interactions. These interactions play an important role in the
chemistry of life and in the properties of both natural and
synthetic compounds. For example, the low molecular mass of
water suggests that it should be a gas at room temperature. The

fact that it is a liquid is due to the hydrogen bonds that exist between individual water molecules. This acts as a weak 'glue' between the molecules and results in a higher boiling point than expected since more energy is required to break the intermolecular hydrogen bonds (Figure 20). Similarly, carboxylic acids have a higher boiling point than might be expected because of hydrogen bonding.

What is hydrogen bonding and how does it occur? Hydrogen bonding relies on the presence of partially charged atoms within a molecule. For example, the oxygen atom in water has a partial negative charge that is indicated with the symbol δ− in Figure 20, while the hydrogen atoms have a partial positive charge (δ+). This partial charge results from the different electronegativities of the oxygen and hydrogen atoms making up a water molecule. Oxygen is to the right of the periodic table, and so it is more electronegative than hydrogen. As a result, oxygen has a greater

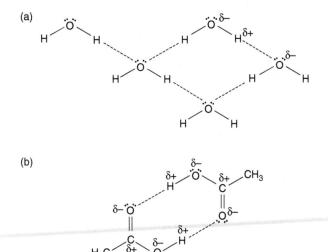

20. Hydrogen bonding between (a) water molecules and (b) carboxylic acids.

pull on the electrons within each O–H bond. Since the electrons in the bond end up closer to oxygen, oxygen becomes slightly negative and hydrogen becomes slightly positive. Thus, the O–H bonds making up water are polar covalent in nature, rather than covalent.

Because of these partial charges, different molecules of water can interact with each other such that a partially charged oxygen atom in one molecule interacts with a partially charged hydrogen atom in another molecule. This is indicated by the dashed lines in Figure 20. Since the interaction is between a slightly negative charge and a slightly positive charge, the interaction can be viewed as a weak form of ionic interaction. However, the interaction is known as hydrogen bonding because a slightly positive hydrogen atom is involved. The hydrogen atom involved in the hydrogen bond is known as the hydrogen bond donor (HBD), whereas the slightly negative oxygen atom is known as the hydrogen bond acceptor (HBA).

Hydrogen bonding between molecules can occur whenever one molecule contains an electronegative atom with a partial negative charge (the HBA), and the other molecule contains a hydrogen atom with a partial positive charge (the HBD). Typically, the HBA is oxygen or nitrogen, while the HBD is a hydrogen atom linked to oxygen or nitrogen. Hydrogen bonding plays an important role in molecular recognition, such as the ability of an enzyme to recognize a substrate (Chapter 4), or the ability of a particular drug or pesticide to bind to a protein target (Chapters 5 and 6).

Intermolecular ionic interactions are possible if one molecule has a positively charged functional group, and the other molecule has a negatively charged functional group. This can occur if one molecule has an aminium group, and the other molecule has a carboxylate group (Figure 21). An ionic interaction is much stronger than a hydrogen bond.

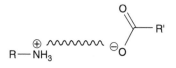

Aminium group   Carboxylate group

21. An ionic interaction between two opposite charges on different molecules.

London dispersion forces (or van der Waals interactions) typically occur between hydrocarbon regions of different molecules; in other words regions containing only carbon and hydrogen atoms. This interaction is much weaker than a hydrogen bond or an ionic bond, but should not be underestimated. There are often more van der Waals interactions between molecules than hydrogen bonds or ionic interactions, and so the cumulative effect of these interactions can be very significant. The interaction is possible because of the random movement of electrons around atoms and molecules. This can result in regions which are briefly electron rich or electron deficient. For any specific region, the effect is only brief and transient. Nevertheless, these transient regions of variable electron density can result in an attractive interaction between molecules, where a transient electron-rich region in one molecule interacts with a transient electron-deficient region in another.

Hydrogen bonding, ionic interactions, and dispersion forces can also take place between different regions of the same molecule. When this occurs, the interactions are defined as intramolecular rather than intermolecular. Such interactions play an important role in the way large molecules (macromolecules) such as proteins and nucleic acids fold up into particular shapes.

Chapter 3

The synthesis and analysis of organic compounds

The design of novel medicines, insecticides, perfumes, flavourings, or polymeric materials relies crucially on organic chemists. That is because organic chemists are specialists in exploring the molecular world, and are trained to understand the structure, properties, and reactions of organic molecules. Moreover, organic chemists have the practical skills to synthesize novel structures, which makes research in an organic synthetic laboratory both challenging and inspiring. Synthetic research is never routine and each day can be a voyage of discovery. Organic synthesis is never totally predictable, and a reaction may well produce a different product from the one planned. This can be frustrating at times, but it can also provide new opportunities for research if that product proves to have useful properties.

Organic research is both creative and practical. Creativity is required in order to design novel molecules that are predicted to have useful properties. It is also required when devising a synthetic route to a specific compound—something akin to chemical chess. To achieve these two goals, the researcher has to have a deep theoretical understanding of organic chemistry, as well as the capability to apply that knowledge to new problems in an imaginative way. However, research chemists also need finely tuned practical skills if they are to carry out synthetic procedures in the laboratory. A good researcher has the chemical equivalent of a gardener's 'green fingers'. Some organic chemists appear to have a magic touch, and are able to carry out reactions more successfully than others. An organic chemist also has to have good analytical skills in order to prove that the product obtained from a reaction is the one intended. If the product proves to be something else, the researcher takes on the role of chemical detective in identifying the structure, and working out how it was formed.

Devising a synthesis

The synthesis of an organic compound must ensure that each atom is in the correct position within the molecule. This could be compared to building a cathedral where every stone is accurately cemented into its correct position. However, this is rather a poor analogy. A cathedral is built stone by stone, whereas it is impossible to build a molecule atom by atom. Instead, target molecules are built by linking up smaller molecules. These smaller molecules (starting materials) must be commercially available, and should ideally resemble part of the target molecule. For example, mepivacaine is a local anaesthetic which can be easily synthesized from two commercially available molecules that resemble each half of the structure (Figure 22). In order to link the two molecules, a reaction is carried out between a functional group on one molecule and a functional group on the other molecule. In this case, one molecule contains an amine, while the other contains an ester. The reaction of an ester with an amine results in the amide required in the target molecule.

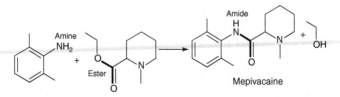

22. Synthesis of mepivacaine.

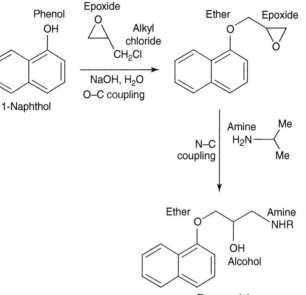

23. Synthesis of propranolol.

The organic chemist needs to have a good understanding of the reactions that are possible between different functional groups when choosing the molecular building blocks to be used for a synthesis. In addition, it is just as important to know when a reaction is unlikely to take place. This can be demonstrated with the synthesis of propranolol—a beta-blocker that is used to treat high blood pressure (Figure 23).

This is a two-step synthesis involving three molecular building blocks. The first step involves the reaction of 1-naphthol with a molecule that contains two functional groups—an epoxide and an alkyl chloride. Under basic conditions, the phenol group of 1-naphthol reacts with the alkyl chloride and displaces the chlorine. This couples the two molecules by means of a new O–C

27

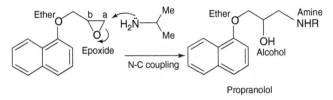

24. A regioselective reaction. The significance of the curly arrows is explained at the end of this chapter.

bond to form a product that contains an ether and the epoxide. Note that the phenol reacts with the alkyl chloride and not the epoxide. This is an example of chemoselectivity where a reaction shows selectivity for one functional group (the alkyl halide) over another (the epoxide).

The product from this reaction is now reacted with the third building block which contains an amine (Figures 23 and 24). This is another chemoselective reaction where the amine reacts with the epoxide rather than the ether. As a result, the three-membered ring of the epoxide is opened up to form an alcohol group and the two molecules are coupled by the

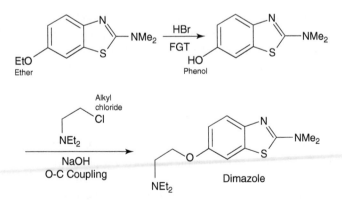

25. Synthesis of dimazole.

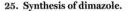

formation of a new N–C bond. This second step has an additional form of selectivity since the amine reacts with the less-substituted carbon of the epoxide ring (position a). This type of selectivity is known as regioselectivity.

The reactions in Figures 22–4 can be described as coupling reactions as they link up different molecular building blocks, but not all the reactions in a synthetic route are of this nature. Indeed, they are often outnumbered by reactions that are described as functional group transformations (FGTs). As the name indicates, these reactions involve converting one functional group to another. There are several reasons why FGTs may be necessary. For example, it may not be possible to obtain a molecular building block that contains the functional group required to couple two molecules together. This can be illustrated in the synthesis of an antifungal agent called dimazole (Figure 25).

The first building block in this synthesis contains an ether functional group. However, this is quite an unreactive functional group and so a coupling reaction is not possible. Therefore, the first reaction in this two-stage synthesis involves a functional group transformation where the ether is converted to a more reactive phenol group by treating it with hydrogen bromide (HBr). The second step of the synthesis is a coupling reaction where the phenol group reacts with an alkyl chloride in the second molecular building block.

A functional group transformation also comes in useful in the synthesis of procaine (a local anaesthetic) (Figure 26). In this case, the FGT is the final step of the synthesis and involves the conversion of a nitro group to an amine. An amine group is quite a reactive group and would interfere with the first two coupling reactions leading to unwanted side products. Therefore, it is 'disguised' as the less reactive nitro group for these two reactions.

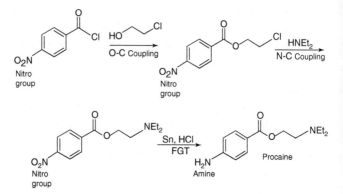

26. Synthesis of procaine.

There are many other reasons for carrying out FGTs, especially when synthesizing complex molecules. For example, a starting material or a synthetic intermediate may lack a functional group at a key position of the molecular structure. Several reactions may then be required to introduce that functional group.

On other occasions, a functional group may be added to a particular position then removed at a later stage. One reason for adding such a functional group would be to block an unwanted reaction at that position of the molecule.

Another common situation is where a reactive functional group is converted to a less reactive functional group such that it does not interfere with a subsequent reaction. Later on, the original functional group is restored by another functional group transformation. This is known as a protection/deprotection strategy.

The more complex the target molecule, the greater the synthetic challenge. Complexity is related to the number of rings, functional groups, substituents, and chiral centres that are present. For example, the local anaesthetic benzocaine has a much simpler

Benzocaine
Molecular mass 165
1 ring, 0 chiral centres
1 substituent,
2 functional groups

Morphine
Molecular mass 285
5 rings, 5 chiral centres
1 substituent,
6 functional groups

27. Comparison of molecular complexity for benzocaine with morphine.

structure than the analgesic morphine (Figure 27). Benzocaine
can be synthesized in one reaction using two molecular building
blocks, whereas the first synthesis of morphine required a total of
twenty-nine reactions. The more reactions that are involved in a
synthetic route, the lower the overall yield. For example, the
twenty-nine-step synthesis of morphine was achieved with an
overall yield of only 0.0014 per cent. Moreover, the final product

was racemic. In other words, it contained a mixture of both enantiomers (mirror images) of this chiral molecule. This meant that only half of the product corresponded to the naturally occurring enantiomer. With such low yields, it is more economic to extract morphine from the poppy plant than to carry out a full synthesis.

It takes great skill and creativity to design and carry out multi-step syntheses of complex molecules successfully, and the chemists involved require a thorough knowledge of the reactions that are possible for different functional groups. Consequently, a number of organic chemists have been awarded the Nobel Prize in Chemistry for their efforts in developing the synthesis of complex natural products from simple starting materials. For example, Sir Robert Robinson was a British organic chemist who gained the Nobel Prize in 1947 for devising the synthesis of various alkaloids, while the American chemist Robert Woodward was honoured in 1965 for devising syntheses of complex natural products such as quinine, cholesterol, strychnine, and chlorophyll. E. J. Corey is another renowned American chemist who won the Nobel Prize in 1990 for synthesizing complex molecules and developing new synthetic methods.

One of the biggest current challenges is the synthesis of maitotoxin which is a high molecular mass, multicyclic neurotoxin produced by a species of plankton in the Pacific Ocean. A number of food poisoning cases have resulted from eating fish that have consumed this plankton. The total synthesis of such a complex molecule will never be a commercial venture, but synthesizing simpler fragments of the molecule might lead to the discovery of new drugs that could be used to treat neurodegenerative diseases. This might seem a bizarre suggestion. If maitotoxin is toxic, then simpler fragments of the structure may also be toxic. However, just because a compound is toxic does not rule out the possibility of it being used in medicine. There are plenty of examples where poisons or toxins have proved useful as medicines. For example,

Bicyclo[1.1.0]butane   Cubane   Prismane   Dodecahedrane

28. Examples of molecules that were synthesized out of curiosity or a sense of challenge.

the arrow poison tubocurarine, once used by South American tribes for hunting game, has been used as a neuromuscular blocker in surgery. One of the basic principles of medicine is the fact that it is the dose of a drug that largely determines whether it acts as a poison or a therapeutic agent. One of the earliest examples of this was morphine, which was used as a painkiller during the American Civil War. Used in the correct dosage, it proved effective, but if the dose was increased tenfold, it proved fatal.

Not all synthetic research is aimed at a designing and synthesizing organic molecules that have a particular purpose. Sometimes, research is carried out because of the challenge involved. For example, some research groups investigate whether unusual-looking molecules are synthetically feasible (Figure 28). Other research teams set themselves the challenge of synthesizing molecules that are seen as having an aesthetic appeal. For example, some chemists find beauty in the molecular architecture of the buckyball (Chapter 9), while others are drawn to the challenge of synthesizing interlocking molecules called rotaxanes (Chapter 9). This kind of research is often inspired by scientific curiosity, rather than commercial motivation. However, there are certainly potential applications for buckyballs and rotaxanes.

Retrosynthesis

The term retrosynthesis is a bit deceptive and one might be tempted to think that it refers to the carrying out of an old-fashioned synthesis. In fact, retrosynthesis is a strategy by which organic chemists design a synthesis before carrying it out in practice. It is called retrosynthesis because the design process involves studying the target structure and working backwards to identify how that molecule could be synthesized from simpler starting materials. Therefore, a key stage in retrosynthesis is identifying a bond that can be 'disconnected' to create those simpler molecules. Note that the disconnection is not an actual reaction and is purely a planning strategy.

There are several guidelines to help the chemist decide on suitable disconnections, but a key principle is that the molecules identified from the disconnection must be realistic. Moreover, there has to be a known reaction that will allow those molecules to be linked together in the actual synthesis to form that same bond. For that reason, favoured bonds for disconnection include C–O and C–N bonds, because it is possible to create these bonds in good yield using well-known reactions.

As an example, consider the structure shown in Figure 29. A suitable disconnection involves the C–N bond indicated by the squiggly line. A special arrow is used for the disconnection to indicate that this is retrosynthesis and not an actual reaction. The two structures that result from this disconnection are known as synthons and are given opposite charges. Synthons are unlikely to be realistic structures because they would be too reactive. Therefore the next stage is to identify real molecules that would resemble them and to identify whether these molecules would

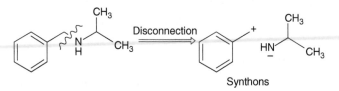

29. Retrosynthesis.

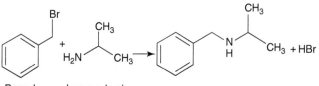

Benzyl
bromide

Isopropylamine

30. The synthesis corresponding to the retrosynthesis shown in Figure 29.

react together to give the desired product. In this case, benzyl bromide and isopropylamine would be suitable starting materials, and could be coupled together as shown in Figure 30.

If the two molecules identified from a disconnection are commercially available, then these molecules can be purchased and the reaction carried out. If the two molecules are not commercially available, then further retrosynthetic analysis is carried out until available starting materials *are* identified. With complex target structures, a retrosynthetic scheme will involve several steps, matching the number of steps that will be needed in the synthesis itself.

Carrying out and monitoring a reaction

Carrying out a reaction is fundamentally very simple. Compound A is mixed with compound B, usually with a solvent that dissolves both compounds. Water would be the ideal solvent in terms of cost, safety, and minimal environmental impact. Unfortunately, most organic compounds are insoluble in water, and so an organic solvent is more commonly used. Commonly used solvents include ethanol, dichloromethane, tetrahydrofuran, ethyl ethanoate (ethyl acetate), propanone (acetone), toluene, dimethyl sulphoxide, and dimethylformamide. Each solvent has its advantages and disadvantages, and choosing the best one for a particular reaction is often down to what has worked best in the past.

Having mixed the various components of a reaction together, it is important to monitor how it proceeds. The popular image of chemical reactions is of instant colour changes, lots of fizzing and steaming, and the occasional bang. In reality, very few reactions produce the spectacular visual and audible effects observed in chemistry demonstrations. More typically, reactions involve mixing together two colourless solutions to produce another colourless solution. Temperature changes are a bit more informative. If heat is generated by the reaction (an exothermic reaction) then the reaction solution increases in temperature. However, not all reactions generate heat, and monitoring the temperature is not a reliable way of telling whether the reaction has gone to completion or not. A better approach is to take small samples of the reaction solution at various times and to test these by chromatography or spectroscopy.

There are various ways of monitoring a reaction by chromatography. One of the simplest is called thin-layer chromatography (TLC), which involves the use of a glass or plastic plate coated with a thin layer of silica (Figure 31). A sample of the reaction solution is spotted on to the silica near the bottom of the plate, alongside samples of both starting materials. The solvent is allowed to

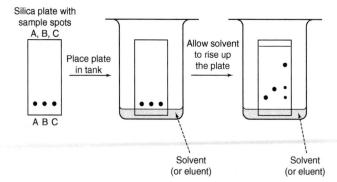

31. Thin-layer chromatography. A and B are the two starting materials for a reaction. C is the reaction mixture.

evaporate, leaving the compounds as dried spots on the plate. The plate is then placed in a chromatographic tank containing another solvent.

This solvent now rises up the plate by capillary action and, as it does so, it drags the compounds along with it. Different compounds will move to different extents up the plate depending on how polar they are. The more polar the compound, the less distance it travels up the plate. This is because silica is a polar material, which means that polar compounds will 'stick' to it more than non-polar compounds. Once the solvent has nearly reached the top of the plate, the plate is removed from the chromatographic tank and the solvent is allowed to evaporate. The spots on the TLC plate can be easily seen if the compounds involved are inherently coloured. Unfortunately, most compounds are colourless and so the plate has to be stained in order to reveal where the spots are. One method is to treat the TLC plate with fumes of iodine. The iodine reacts with any compounds present and shows them up as brown spots.

Using TLC, a reaction can be monitored with time. Early on, very little reaction will have taken place, in which case the sample (C) from the reaction mixture will contain mostly starting materials A and B (Figure 32). As the reaction proceeds, a new spot corresponding to the product appears and increases in intensity, while the spots corresponding to compounds A and B diminish in intensity. The reaction can be judged to be complete when no more starting material is present, or when there is no further evidence of change. Note that not all reactions go to completion.

A different approach to monitoring a reaction is to test samples of the reaction mixture by infrared spectroscopy. This involves subjecting molecules to infrared radiation, and measuring whether any of the radiation is absorbed. Infrared radiation interacts with the bonds in a molecule and causes them to

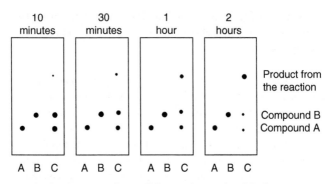

10 minutes	30 minutes	1 hour	2 hours	
				Product from the reaction
				Compound B
				Compound A
A B C	A B C	A B C	A B C	

32. Monitoring a reaction at different times using thin-layer chromatography.

Organic Chemistry

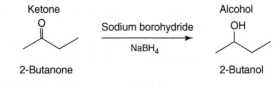

Ketone → Alcohol

Sodium borohydride
$NaBH_4$

2-Butanone → 2-Butanol

33. Reduction of a ketone to an alcohol.

vibrate at characteristic frequencies. When this happens, energy is absorbed. The frequency of absorbed infrared radiation is characteristic for the particular types of functional groups that are present. For example, a carbonyl group (C=O) absorbs a different frequency of infrared radiation from an alcohol (O–H) functional group. Therefore, if a reaction is being carried out where a ketone is being reduced to an alcohol (Figure 33), the reaction can be monitored by following the rate at which the carbonyl absorption diminishes and the hydroxyl absorption increases. For example, in the reduction of 2-butanone to 2-butanol, the carbonyl absorption of 2-butanone at 1,715 cm^{-1} would gradually diminish, while the hydroxyl absorption of 2-butanol at 3,350 cm^{-1} would gradually increase (Figure 34).

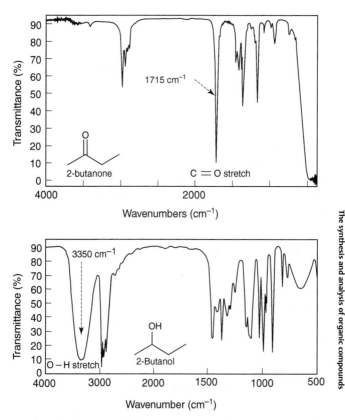

34. The infrared spectra for 2-butanone and 2-butanol.

Changing the conditions of a reaction

If a reaction is taking place very slowly, different reaction conditions could be tried to speed it up. This could involve heating the reaction, carrying out the reaction under pressure, stirring the contents vigorously, ensuring that the reaction is carried out in a dry atmosphere, using a different solvent, using a catalyst, or using one of the reagents in excess.

On the other hand, a reaction might be too vigorous and result in the appearance of unwanted side-products and impurities. In some reactions, the product might be formed, but then degrade or undergo further reactions. Again, altering the reaction conditions might improve the situation. For example, the reaction could be carried out at a cold temperature, or under a nitrogen atmosphere.

There are a large number of variables that can affect how efficiently reactions occur, and organic chemists in industry are often employed to develop the ideal conditions for a specific reaction. This is an area of organic chemistry known as chemical development.

Isolation and purification of a reaction product

Once a reaction has been carried out, it is necessary to isolate and purify the reaction product. This often proves more time-consuming than carrying out the reaction itself. Ideally, one would remove the solvent used in the reaction and be left with the product. However, in most reactions this is not possible as other compounds are likely to be present in the reaction mixture. For example, the reaction may not have gone fully to completion, in which case small quantities of starting materials and reagents will still be present. This is particularly the case when an excess of one starting material has been added in order to get a good yield. Inorganic salts may also have been formed depending on the type of reagents used. Finally, there may be impurities present as a result of side reactions, where some of the starting material has undergone a different reaction from the one intended. Therefore, it is usually necessary to carry out procedures that will separate and isolate the desired product from these other compounds. This is known as 'working up' the reaction.

A typical reaction work up starts with various extraction procedures. If the reaction has been carried out in an organic solvent that does not mix with water then the extraction process can be carried out straight away. Examples of such solvents are dichloromethane, ethyl acetate, and diethyl ether. If the reaction was carried out in a solvent that *does* mix with water, then the solvent has to be removed by evaporation. The crude reaction product is then dissolved in a suitable organic solvent that will not mix with water.

Once the crude reaction mixture is dissolved in a suitable solvent, the extraction process can be carried out (Figure 35). As an illustration, we shall assume that the crude reaction mixture contains four different organic compounds (a–d). Compound (d) is the desired product, compounds (a) and (c) are unreacted starting materials, where compound (a) is an amine and compound (c) is a carboxylic acid. Compound (b) is an impurity that has been formed in the reaction as a side reaction.

The solution of the mixture is first poured into a separating funnel. An aqueous solution of sodium hydroxide is then added to give two immiscible phases. The separating funnel is stoppered and shaken to mix the phases, then the phases are allowed to separate back out again. By doing this, the sodium hydroxide solution ionizes any carboxylic acid (c) that is present in the mixture to produce a water-soluble carboxylate ion (Figure 19). This moves into the water layer. The phases are now separated, with compound (c) ending up in the aqueous layer.

The organic layer is returned to the separating funnel and shaken with a solution of hydrochloric acid (HCl), which will ionize any amine that is present (Figure 19). Again, the two phases are allowed to settle, then separated. The ionized amine is water soluble and ends up in the HCl solution.

The organic layer now contains the impurity (b) and the product (d). The fact that the impurity was not extracted into the basic or

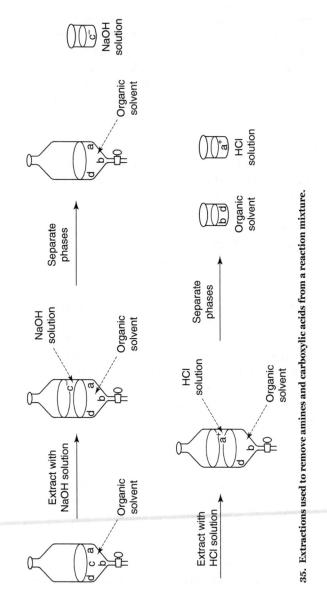

35. Extractions used to remove amines and carboxylic acids from a reaction mixture.

acidic solution indicates that it does not contain an acidic or basic functional group.

The next stage is to dry the organic layer in order to remove any traces of water. This is done by adding an anhydrous salt, such as magnesium sulphate, which soaks up the water without dissolving in the organic solvent. The salt can be removed by filtration, and the organic solvent is then removed by distillation to give the crude product (d), contaminated with the impurity (b).

Purification of the crude product is now required in order to remove the impurity (b). One possible method is to carry out a crystallization. This involves dissolving up the crude product into a solvent in which product (d) is only sparingly soluble. This means that heating is required in order to get the product to dissolve. The hot solution is then left to cool. As it does so, product (d) crystallizes out of solution. Assuming that the impurity is present in a much smaller quantity, it is less likely to crystallize and will stay in solution. The crystals of the pure product can then be filtered off.

Unfortunately, many organic compounds do not crystallize particularly well, or are obtained as oils. In such cases, a common approach is to separate the product from the impurity using chromatography. The principle is the same as TLC (Figure 31) but carried out on a larger scale, using silica packed into glass columns. A solution of the crude mixture is then added to the top of the column and the different products descend through the silica at different rates as solvent is passed down the column. Once each compound has passed through the silica, it is collected at the bottom and the solvent is distilled off to give the pure compound.

Structural analysis

Assuming that a reaction has been successful and a product has been isolated and purified, it is essential to establish the structure of that product. The outcome of a reaction is never totally predictable, and there is always the possibility that a reaction may produce a different product from the one intended. Organic synthesis is not like a civil engineering project where girders are linked together in a predictable way. Molecules can react in surprising or unexpected ways.

Ideally, it would be nice to look at the product under a microscope and see the molecule directly. However, this is not possible. The closest one can come to visualizing a structure directly is to obtain a crystal of the product and to carry out a technique called X-ray crystallography. This can determine the atoms that are present and provide a visual representation of how those atoms are linked together—the closest that one can get to a molecular photograph. However, X-ray crystallography takes a relatively long time. Moreover, a large proportion of organic molecules synthesized in the laboratory are oils or liquids, or fail to give satisfactory crystals. Other, less direct methods of determining structure are required for such compounds.

There are a number of analytical tools that can be used to determine structure. For example, it is possible to identify what elements are present in a compound, and their relative proportions, using a procedure known as elemental analysis. It is also possible to determine the molecular mass using mass spectrometry. Together, these two analytical methods provide the molecular formula for a compound. However, they do not reveal how the different atoms are linked together. As we have already seen (Figure 34), infrared spectroscopy can be used to determine whether specific functional groups are present, but the spectrum provides no information about the carbon skeleton of the molecule.

For many years, there was no unambiguous way of determining the structure of a molecule other than by X-ray crystallography, and so determining the structure of many compounds took several years to achieve. From the 1960s, this all changed with the advent of a technique called nuclear magnetic resonance (NMR) spectroscopy. NMR spectroscopy uses electromagnetic radiation to detect the nuclei of specific elements or isotopes. The most common form of NMR is ^{1}H or proton NMR, which detects the nuclei of all the hydrogen atoms present in a structure. Another common form of NMR is carbon NMR, which detects all the carbon nuclei in a molecular structure. In an NMR spectrum, each relevant atom in the structure is revealed as a signal, and the position of that signal is indicative of the molecular environment of the atom. This helps to identify where each atom is located in a molecule. The carbon NMR spectra for 2-butanone and 2-butanol (Figure 36) show four signals for the four carbon atoms present in each structure. The biggest difference between the two spectra is the signal due to carbon-b.

Proton NMR spectra are more complex because some signals are split into several peaks. This is known as the coupling pattern. For example, the proton spectrum for 2-butanone (Figure 37) has three signals representing the hydrogens attached to carbons a, c, and d. Signal a is a single peak (a singlet) for the three hydrogen atoms linked to carbon a. Signal c has four peaks (a quartet) for the two protons attached to carbon c, and signal d has three peaks (a triplet) for the three protons attached to carbon d. There is no signal b because there are no hydrogen atoms attached to that carbon.

The coupling patterns for signals c and d may complicate the spectrum, but they provide very useful information about the molecular structure. This is because the number of peaks within any signal indicates the number of protons that are present on a neighbouring carbon. To be specific, the number of peaks observed in a signal is one more than the number of protons on a neighbouring carbon. For example, signal c (the CH_2 group) has four peaks, which indicates that there are three protons on a

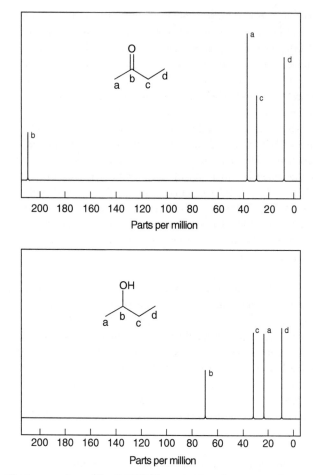

36. A comparison of the C-13 NMR spectra for 2-butanone and 2-butanol.

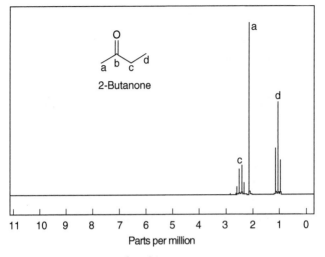

37. A proton NMR spectra for 2-butanone.

neighbouring carbon (the methyl group at position d). Similarly, signal d (the CH_3 group) has three peaks, which indicates the presence of two protons on a neighbouring carbon (the CH_2 group at position c). Together, this proves the presence of an ethyl group in the structure (CH_2CH_3). NMR analysis such as this allows chemists to work out how atoms are linked to each other throughout the whole structure.

Reaction mechanisms

An important part of organic chemistry is understanding how reactions take place. Reactions involve the making and breaking of covalent bonds. The mechanism for a particular reaction identifies the electrons involved. In other words, if a new bond is formed, where did the two electrons for the new bond come from? Alternatively, if a bond is broken, where did the two electrons making up that bond go?

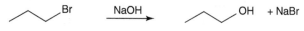

1-Bromopropane   1-Propanol

38. Reaction of 1-bromopropane with 1-propanol.

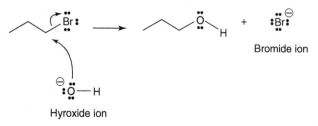

Bromide ion

Hyroxide ion

39. Mechanism for the reaction shown in Figure 38.

To illustrate this, consider the reaction shown in Figure 38. This involves the reaction of 1-bromopropane with sodium hydroxide to form 1-propanol. The reaction involves the breaking of a bond between carbon and bromine, and the formation of a bond between carbon and the oxygen atom of the hydroxide group (OH).

The mechanism for this reaction is shown in Figure 39. The curly arrows indicate the movement of pairs of electrons in order to make and break bonds. For example, the curly arrow at the top indicates that the two electrons in the C-Br bond are moving to bromine. As a result, the bond between carbon and bromine is broken and the two electrons involved are now present on the bromide ion as a fourth lone pair of electrons. The bromide ion also gains a negative charge as a result.

The bottom curly arrow shows that a lone pair of electrons on the oxygen atom of the hydroxide ion is being used to form a new bond between oxygen and carbon. As a result, the oxygen atom in the product now has two lone pairs of electrons instead of three. It also loses its negative charge.

This is a relatively simple mechanism, but it illustrates the principle behind the use of curly arrows. The curly arrow must start from a pair of electrons, which means that it can only be drawn from a lone pair of electrons on an atom, or a covalent bond between two atoms. The arrow must point to where the electrons end up. This can either be a new covalent bond between two atoms or a new lone pair on an atom.

Chapter 4
The chemistry of life

It was once believed that organic compounds were unique to life and could not be synthesized in the laboratory. We now know that this is not the case, but it is clear that organic molecules are fundamental to the existence of life on this planet. Chapter 3 described how organic chemists synthesize organic molecules from smaller molecular building blocks. This might seem very clever, but nature has been doing the same thing for much longer, and does it far more efficiently. From very simple molecular building blocks, life has created an astonishing diversity of molecules, some of which are extremely complex structures that prove very difficult to synthesize in a laboratory. Not only does life produce complex molecules, it does so under mild conditions in an aqueous environment.

Amino acids and proteins

Proteins are large molecules (macromolecules) which serve a myriad of purposes, and are essentially polymers constructed from molecular building blocks called amino acids (Figure 40). In humans, there are twenty different amino acids having the same 'head group', consisting of a carboxylic acid and an amine attached to the same carbon atom (Figure 41). The simplest amino acid is glycine which has a hydrogen atom in place of a side chain, but all the other amino acids have some form of side chain present.

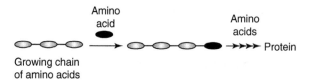

Figure 40

40. Biosynthesis of proteins formed by linking amino acids one at a time.

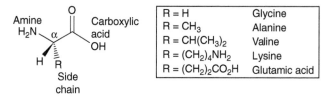

R = H	Glycine
R = CH$_3$	Alanine
R = CH(CH$_3$)$_2$	Valine
R = (CH$_2$)$_4$NH$_2$	Lysine
R = (CH$_2$)$_2$CO$_2$H	Glutamic acid

41. Structures of selected α-amino acids.

The amino acids are linked up by the carboxylic acid of one amino acid reacting with the amine group of another to form an amide link. Since a protein is being produced, the amide bond is called a peptide bond, and the final protein consists of a polypeptide chain (or backbone) with different side chains 'hanging off' the chain (Figure 42). The sequence of amino acids present in the polypeptide sequence is known as the primary structure. Once formed, a protein folds into a specific 3D shape, which is determined by the intramolecular interactions occurring between different side chains and peptide bonds, as well as intermolecular hydrogen bonding to surrounding water molecules. Life's method of synthesizing proteins can be mimicked in the laboratory. For

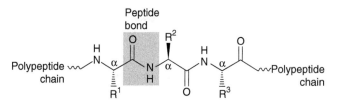

42. The polypeptide backbone of a protein with substituents R^1, R^2, R^3, etc.) attached to each α-carbon.

example, HIV protease is a viral enzyme that has been synthesized in the laboratory.

Nucleotides and nucleic acids

Nucleic acids (Figure 43) are another form of biopolymer, and are formed from molecular building blocks called nucleotides. These link up to form a polymer chain where the backbone consists of alternating sugar and phosphate groups. There are two forms of

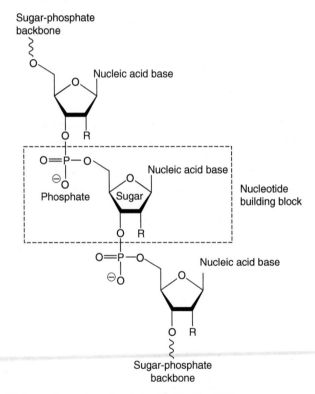

43. General structure of nucleic acids (R=H or OH).

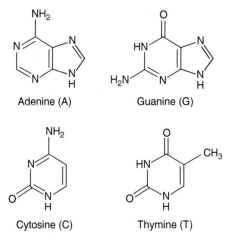

Adenine (A)

Guanine (G)

Cytosine (C)

Thymine (T)

44. The nucleic acid bases present in DNA.

nucleic acid—deoxyribonucleic acid (DNA) and ribonucleic acid (RNA). In DNA, the sugar is deoxyribose (R=H), whereas the sugar in RNA is ribose (R=OH). Each sugar ring has a nucleic acid base attached to it. For DNA, there are four different nucleic acid bases called adenine (A), thymine (T), cytosine (C), and guanine (G) (Figure 44). These bases play a crucial role in the overall structure and function of nucleic acids.

DNA is actually made up of two DNA strands (Figure 45) where the sugar-phosphate backbones are intertwined to form a double helix. The nucleic acid bases point into the centre of the helix, and each nucleic acid base 'pairs up' with a nucleic acid base on the opposite strand through hydrogen bonding. The base pairing is specifically between adenine and thymine, or between cytosine and guanine. This means that one polymer strand is complementary to the other, a feature that is crucial to DNA's function as the storage molecule for genetic information.

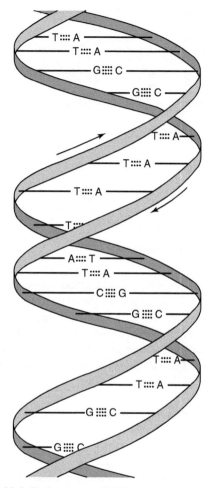

45. The double helical structure of DNA.

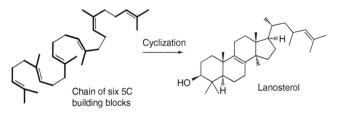

Cyclization

Chain of six 5C
building blocks

HO

Lanosterol

46. The process involved in steroid biosynthesis.

Other biosynthetic processes

Polymerization processes are involved in the biosynthesis of
many other natural products. For example, fatty acids are formed
from 2-C building blocks. In the case of cyclic compounds, a
polymerization process creates a linear polymer chain, which then
undergoes cyclization reactions. For example, the biosynthesis of
steroids creates a polymer chain made up of 5-C building blocks,
which is cyclized to form a tetracyclic structure (Figure 46). The
same general strategy is used in different organisms. For example,
penicillin G is a fungal metabolite that is created from linking up
two amino acids and one fatty acid, then carrying out cyclization
reactions (Figure 47).

Function of proteins

Proteins have a variety of functions. Some proteins, such as
collagen, keratin, and elastin, have a structural role. Others
catalyse life's chemical reactions and are called enzymes. They
have a complex 3D shape, which includes a cavity called the active
site (Figure 48). This is where the enzyme binds the molecules
(substrates) that undergo the enzyme-catalysed reaction. The
resulting product is then released.

A substrate has to have the correct shape to fit an enzyme's active
site, but it also needs binding groups to interact with that site
(Figure 49). These interactions hold the substrate in the active site

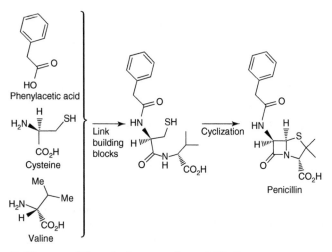

47. **The general biosynthetic process for penicillin G.**

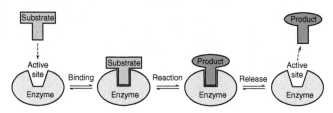

48. **The overall process involved in an enzyme-catalysed reaction.**

long enough for a reaction to occur, and typically involve hydrogen bonds, as well as van der Waals and ionic interactions. When a substrate binds, the enzyme normally undergoes an induced fit. In other words, the shape of the active site changes slightly to accommodate the substrate, and to hold it as tightly as possible.

Once a substrate is bound to the active site, amino acids in the active site catalyse the subsequent reaction. One example of an enzyme-catalysed reaction is the hydrolysis of a neurotransmitter

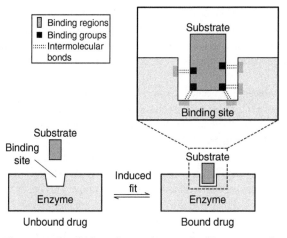

49. The process by which a substrate is recognized by an enzyme's active site.

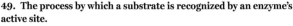

50. The enzyme-catalysed hydrolysis of acetylcholine.

called acetylcholine (Figure 50). In water, this reaction is very slow, but it is several million times faster if the enzyme is present.

Proteins called receptors are involved in chemical communication between cells and respond to chemical messengers called neurotransmitters if they are released from nerves, or hormones if they are released by glands. Most receptors are embedded in the cell membrane, with part of their structure exposed on the outer surface of the cell membrane, and another part exposed on the inner surface. On the outer surface they contain a binding site that binds the molecular messenger. An induced fit then takes place that activates the receptor. This is very similar to what happens when a substrate binds to an enzyme (Figure 49). However, the receptor has no catalytic activity. The molecular messenger stays bound for a certain period of time, then departs unchanged. Once that happens, the protein receptor returns to its inactive state.

The induced fit is crucial to the mechanism by which a receptor conveys a message into the cell—a process known as signal transduction. By changing shape, the protein initiates a series of molecular events that influences the internal chemistry within the cell. For example, some receptors are part of multiprotein complexes called ion channels. When the receptor changes shape, it causes the overall ion channel to change shape. This opens up a central pore allowing ions to flow across the cell membrane. The ion concentration within the cell is altered, and that affects chemical reactions within the cell, which ultimately lead to observable results such as muscle contraction. Not all receptors are membrane-bound. For example, steroid receptors are located within the cell. This means that steroid hormones need to cross the cell membrane in order to reach their target receptors.

Transport proteins are also embedded in cell membranes and are responsible for transporting polar molecules such as amino acids into the cell. They are also important in controlling nerve action since they allow nerves to capture released neurotransmitters, such that they have a limited period of action. Transport proteins contain an extracellular binding site that binds the target molecule. Once bound, the molecule is passed through the protein and released into the cell.

Function of nucleic acids

DNA is the molecule responsible for storing genetic information, and transmitting it from one generation to another. However, for many years, DNA was considered an unlikely candidate for these roles. The backbone was a regular sequence of sugar and phosphate groups, and there were only four different types of nucleic acid base present, which appeared to be randomly arranged along the chain. However, attitudes changed dramatically when it was discovered that DNA was a double helix held together by interactions between specific base pairs. This meant that the two chains were complementary and demonstrated how DNA could pass on genetic information from one generation to another. By unwinding the double helix, each strand could act as the template for the creation of a new strand to produce two identical 'daughter' DNA double helices (Figure 51). But this still left the question of how a genetic alphabet of four letters (A, T, G, C) could code for twenty amino acids. The answer lies in the triplet code, where an amino acid is coded, not by one nucleotide, but by a set of three. The number of possible triplet combinations using four 'letters' is more than enough to encode all the amino acids.

RNA is the other form of nucleic acid present in cells, and is crucial to protein synthesis (translation). There are three forms of RNA—messenger RNA (mRNA), transfer RNA (tRNA), and ribosomal RNA (rRNA). mRNA carries the genetic code for a particular protein from DNA to the site of protein production. Essentially, mRNA is a single-strand copy of a specific section of DNA. The process of copying that information is known as transcription.

tRNA decodes the triplet code on mRNA by acting as a molecular adaptor. At one end of tRNA, there is a set of three bases (the anticodon) that can base pair to a set of three bases on mRNA (the codon). An amino acid is linked to the other end of the tRNA and the type of amino acid present is related to the anticodon that is present. When tRNA with the correct anticodon base pairs to the codon on mRNA, it brings the amino acid encoded by that codon.

rRNA is a major constituent of a structure called a ribosome, which acts as the factory for protein production. The ribosome binds mRNA then coordinates and catalyses the translation

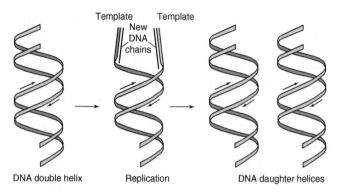

Template **Template**

New
DNA
chains

DNA double helix   Replication   DNA daughter helices

51. Replication of DNA chains.

process. Each ribosome attaches to the end of mRNA, and travels
along its length. As it does so, only two codons are exposed at any
one time allowing the binding of two tRNA molecules to the
ribosome. The protein is built up one amino acid at a time, with
the growing protein chain being transferred from one tRNA to
the amino acid on the other (Figure 52).

Chemical evolution

In the 1880s, Charles Darwin proposed that the organic chemicals
essential for life might have been formed in a 'warm little pond'.
Since then, organic chemists have postulated how the essential
building blocks for life might have been formed on Earth during
the one billion-year period before life began. This is a topic known
as chemical evolution or prebiotic chemistry.

The first chemists to study chemical evolution seriously were
Stanley L. Miller and Harold C. Urey who worked at the
University of Chicago. In 1953, they conducted an experiment that
was designed to mimic the environmental conditions that might
have been present during the prebiotic period. They proposed that
Earth's early atmosphere contained gases such as methane and

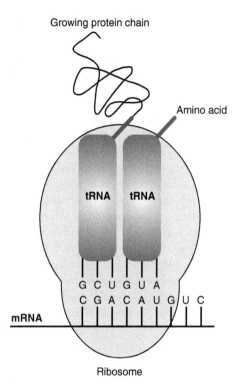

Growing protein chain

Amino acid

tRNA   tRNA

G C U G U A
C G A C A U G U C

mRNA

Ribosome

52. Translation.

ammonia rather than oxygen, which only appeared once plant life
had become established.

The experiment devised by Miller and Urey involved a
round-bottomed flask containing water to mimic the oceans,
and a mixture of methane, ammonia, and hydrogen to represent
the atmosphere. A couple of electrodes were inserted into the
flask, across which electrical charges could be passed to mimic
lightning. It was proposed that lightning could provide the energy
required to drive reactions between the gaseous molecules, and
that the products formed would end up dissolved in the 'ocean'.

The experiment was carried out for a week and, when the water was analysed, a number of naturally occurring amino acids were found to be present. This single experiment convinced many scientists that life could be created, not only on Earth, but elsewhere in the cosmos.

Subsequent experiments demonstrated that similar reactions took place with heat or ultraviolet radiation as the energy source. Experiments were also carried out with different gases in the 'atmosphere', which demonstrated that oxygen prevented the formation of amino acids. If hydrogen cyanide was present, adenine was formed. Adenine is one of the nucleic acid bases, and appears in several biochemicals, suggesting that it may have been formed early on during chemical evolution. Indeed, the structure of adenine is made up entirely of C–N segments that could be derived from five cyanide molecules (Figure 53). When formaldehyde was included in the 'atmosphere', ribose was formed. Many scientists now believe that Earth's prebiotic atmosphere contained carbon dioxide, carbon monoxide, and

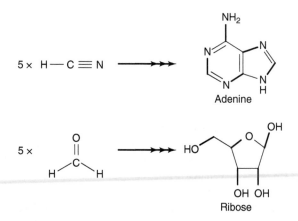

53. Possible building blocks for adenine and ribose.

nitrogen. The Miller–Urey type experiments carried out under this mixture of gases also produces life's molecular building blocks.

An alternative theory of chemical evolution is that the building blocks of life were formed in deep-sea hydrothermal vents. These vents spew forth the mixtures of chemicals required for prebiotic reactions, and also provide heat as an energy source. Moreover, the products formed would be sheltered from the hostile environment on the planet's surface and have a greater chance of forming biopolymers such as proteins and nucleic acids.

Yet another theory proposes that life's building blocks were synthesized in outer space, then brought to Earth by meteorite showers. Astrochemistry is a branch of chemistry that has been investigating such a possibility by searching for extraterrestrial organic molecules. For example, methane has been detected as a major component of the atmospheres of Jupiter, Saturn, Uranus, and Neptune. NASA's Curiosity Rover has been searching for organic molecules on the surface of Mars, and, in November 2014, the European space agency's Rosetta probe landed a robotic detector called Philae on to a meteorite. Initial data revealed organic molecules in the comet's atmosphere. There is also interest in finding out whether organic molecules are present in the atmosphere of Titan (one of Saturn's moons). The Cassini probe has been orbiting the moon since 2004 and it is now known that Titan has an atmosphere containing polyaromatic hydrocarbons. Titan's methane lakes might also have nitrogen-containing organic molecules. However, a lack of oxygen, and Titan's extremely cold temperatures, make the evolution of life unlikely.

On Earth, a variety of experiments have been carried out to mimic conditions in outer space, and to see what synthetic reactions could take place. One possibility is that reactions are occurring on meteorites as a result of the solar wind acting on formamide—a simple organic molecule that has been detected in significant quantities in outer space. Simulation experiments, where a proton beam has been shone on to a mixture of formamide and meteorite dust, have produced a variety of life's molecular building blocks. There is a particular interest in meteorites called chondrites. These are the oldest meteorites in space and it is believed that they may help to catalyse reactions. If such meteorites then 'seed' planets with the chemical building blocks for life, it is possible that life found elsewhere in the universe would have molecular similarities to life on Earth. One objection to the theory of meteorite 'seeding' is whether organic molecules could survive the extreme heat experienced when a meteorite enters a planet's atmosphere.

One of evolution's mysteries is how single enantiomers of amino acids and carbohydrates came to be selected as the preferred building blocks for proteins and nucleic acids. One theory, called the Vester–Ulbricht hypothesis, is based on the fact that electrons have left- and right-handed spins. Polarized radiation composed of either left-handed or right-handed electrons might have degraded one enantiomer of a chiral molecule more quickly than the other. There is some experimental evidence for this. However, it is yet to be demonstrated for amino acids or carbohydrates.

Evolution of life's biopolymers

Amino acids, carbohydrates, and nucleic acid bases can be formed under a variety of prebiotic environmental conditions, but they fail to polymerize into proteins and nucleic acids. For polymerization to take place, some form of catalysis is required. In living cells, DNA synthesis is catalysed by proteins, but the synthesis of those proteins relies on the genetic information provided by DNA. The two synthetic processes are interdependent. Moreover, life depends on genetic information being passed on from one generation to another. How could this have evolved?

One theory is that a molecule was formed under prebiotic conditions that could catalyse its own replication. There have been debates over whether this was an early protein or DNA structure. However, there are problems with both possibilities. Amino acids can certainly link up to form peptides in the presence of a simple catalyst called cyanamide, which can be formed from methane and ammonia under prebiotic conditions. However, there is no control over the order in which the amino acids are linked together, or any mechanism by which a useful peptide could be copied.

As far as DNA is concerned, small DNA chains called oligonucleotides can be formed under prebiotic conditions. Such molecules could potentially act as templates for identical molecules, but they are too short to contain useful genetic information, and there is no prebiotic catalytic mechanism to promote their replication.

Most scientists now believe that the key molecule that sparked the evolution of life was RNA. This is a reasonable proposal considering the central role that RNA now plays in protein synthesis. An early form of RNA may have evolved that acted as a template for the synthesis of identical copies, in which case RNA would have been the original storage molecule for genetic information. It has also been proposed that this early RNA could have catalysed a self-replication process. Recent research has demonstrated that some current-day RNA molecules do indeed have a catalytic property. Such molecules are called ribozymes. Early life might well have relied on ribozymes to serve the functions currently carried out by DNA and proteins. It has even been suggested that the ribosomes in current-day cells might have evolved from an early ribozyme. The central regions of ribosomes from different organisms are relatively similar, which might suggest a common molecular ancestor from 3.8 billion years ago. That ancestor could well have marked the crucial transition from prebiotic chemistry to life itself.

Making use of proteins and nucleic acids

Enzymes and nucleic acids are increasingly being used in commercial applications. For example, enzymes are used routinely to catalyse reactions in the research laboratory, and for a variety of industrial processes involving pharmaceuticals, agrochemicals, and biofuels. In the past, enzymes had to be extracted from natural sources—a process that was both expensive and slow. But nowadays, genetic engineering can incorporate the gene for a key enzyme into the DNA of fast growing microbial cells, allowing the enzyme to be obtained more quickly and in far greater yield. Genetic engineering has also made it possible to modify the amino acids making up an enzyme. Such modified enzymes can prove more effective as catalysts, accept a wider range of substrates, and survive harsher reaction conditions. For example, a modified enzyme was used to catalyse a key synthetic step in the synthesis of sitagliptin—an agent used to treat diabetes. The natural enzyme was unable to catalyse this reaction because the substrate involved was too big to fit the active site. Genetic engineering produced a modified enzyme with a larger active site.

Companies such as Novozymes and Du Pont specialize in designing modified enzymes. For example, biological washing powders contain a variety of enzymes that catalyse the removal of persistent stains caused by food, blood, or sweat. Proteases cleave peptide bonds in proteins, lipases cleave ester groups in lipids and fats, and amylases degrade starches. Cellulases partially degrade the cellulose fibres in clothes to help free trapped dirt particles.

New enzymes are constantly being discovered in the natural world as well as in the laboratory. Fungi and bacteria are particularly rich in enzymes that allow them to degrade organic compounds. It is estimated that a typical bacterial cell contains about 3,000 enzymes, whereas a fungal cell contains 6,000. Considering the variety of bacterial and fungal species in existence, this represents a huge reservoir of new enzymes, and it is estimated that only 3 per cent of them have been investigated so far. Microorganisms that survive extreme climates can be particularly useful sources for enzymes that will operate under harsh conditions. For example, microorganisms that survive in the Arctic have provided enzymes that are effective at room temperature and avoid the need for heating reactions. A protease that acts at high pH was discovered from a microorganism growing in a cemetery. This enzyme has proved a useful addition in detergents.

A number of enzymes including amylase have proved important in the production of biofuels such as ethanol. The enzymes catalyse the breakdown of starch and glycogen found in sugar cane and maize. Unfortunately, there are significant drawbacks to producing biofuels from food crops, since it means that less food reaches the shops and prices increase. A better approach would be to make use of plant material that is currently wasted. Therefore, research is being carried out to find cellulase enzymes that break down the cellulose present in plant leaves and stalks. For example, Novozymes has produced a mixture of cellulase enzymes called Cellic which could be used for bioethanol production.

Enzymes will also prove important in providing many of the reagents and chemicals currently obtained from oil. For example, 5.5 million tons of adipic acid—a key constituent needed in the synthesis of nylon—needs to be produced each year. Oil has been the traditional source, but conventional oil production is predicted to drop by over 50 per cent in the next two decades, and fracking is not a long-term answer. Enzymes will play a crucial role in producing these important chemicals from alternative sources.

There are some unusual applications for enzymes. For example, enzymes are being considered as components in batteries. The idea is to attach enzymes to the electrodes, and to use glucose as a fuel. The enzymes catalyse the oxidation of glucose, producing electrons that are transferred to the electrode. It is predicted that such batteries could eventually power mobile phones, pacemakers, and other small devices.

There are potential applications for other types of protein. For example, a protein called reflectin is largely responsible for the method by which squid, octopus, and cuttlefish camouflage themselves. If the protein is phosphorylated, it causes cells called iridophores to reflect light such that the organism mirrors its surroundings. Researchers are considering using a similar process to design new camouflaged materials. Other experiments are being carried out to see if materials based on this technology could reflect heat. If successful, smart clothing could be designed that regulates how much heat is radiated to the environment, such that a jacket could be cool on warm days and warm on cold days.

Several scientific teams have been investigating proteins that allow organisms to survive in freezing conditions. For example, the larvae of Alaskan beetles can survive temperatures as low as −100°C. It has been found that some proteins can act as a natural antifreeze to prevent ice crystals being formed within cells. The protein structure appears to 'mop up' microcrystals of ice and prevent them growing any larger. There are potential commercial uses for proteins that behave in this way. Indeed, antifreeze proteins are already being used in ice creams because they limit the size of ice crystals that are formed. This produces a smooth texture and allows the fat content to be cut.

A large amount of research is also taking place into applications for DNA. For example, DNA is being considered as a data storage system. Although there are only four 'letters' in DNA, a binary system can be used to store information, and scientists have demonstrated this by storing Martin Luther's speech 'I have a dream' on a DNA system. The number of 'letters' incorporated in DNA molecules could also be increased by including synthetic base pairs. For example, synthetic bases called d5SICS and dNAM have been designed that can base pair in DNA through hydrophobic interactions. DNA is actually a remarkable stable molecule, and bacterial DNA has been recovered from Antarctic ice cores that are millions of years old.

Other commercial uses for DNA include a number of diagnostic devices that can identify metals in water, toxic gases, food spoilage, and the virus responsible for Ebola. Another possible use for DNA is as a template for polymer synthesis. Experiments have shown that DNA can bind molecular adaptors that carry the monomers required for a polymer. Once the adaptors are bound to the DNA template, the polymerization is carried out, and the polymer is released. This method has already been used to prepare polyethylene glycol.

DNA from herring sperm has been found to have a possible use as a fire retardant. When DNA is degraded by fire, it produces ammonia which prevents oxygen from fuelling the flames. The flame retardant industry is big business. However, many traditional fire retardants are halogenated molecules that are hazardous to the environment, and can prove toxic to both animals and humans. If DNA proves effective as a fire retardant, it would have the advantages of being abundant, naturally occurring, and biodegradable.

DNA even has potential in the development of lithium-sulphur batteries, which consist of a lithium metal anode and a carbon-sulphur cathode. When the battery is active, lithium ions are released from the anode and react with the sulphur at the cathode to form polysulphides. The reaction is reversed when the battery is recharged. However, some of the sulphur is lost from the cathode when the battery is in operation, and this results in a loss of performance. The nucleic acid bases and phosphate groups of DNA have a strong affinity for sulphur and it is thought that coating the cathode with DNA might prevent the loss of sulphur into the electrolyte.

Chapter 5
Pharmaceuticals and medicinal chemistry

One of the most important applications of organic chemistry involves the design and synthesis of pharmaceutical agents—a topic that is defined as medicinal chemistry. This is a relatively new scientific discipline. Before the 1960s, the discovery of pharmaceutical agents was very much a hit-and-miss affair. Thousands of organic compounds were made in the laboratory, or were extracted from natural sources, in the hope that they would have pharmacological activity. Success was more down to luck than design. From the 1960s on, there has been much greater understanding of how drugs work, and of the targets with which they interact. Advances in biology, genetics, chemistry, and computing have now made it possible to design new drugs, rather than to rely on trial and error. Medicinal chemists are key players in the pharmaceutical industry because they are experts at both designing and synthesizing drugs. The two skills go hand in hand. For example, there is no point in designing a drug that cannot be synthesized. Similarly, it is wasteful to generate thousands of novel compounds if they have little chance of being active drugs.

The pathfinder years

In the past, societies depended on herbs and extracts from natural sources to treat illness. No doubt there was a powerful placebo effect involved, as most ancient therapies have little beneficial effect. However, some of these concoctions *are* effective. Examples include the sedative effects of various opium preparations, and the physical and psychological effects obtained from chewing coca leaves—a habit which is still common in some South American communities. Extracts of willow bark have been known for centuries to reduce fever, pain, and inflammation.

In the 19th century, chemists isolated chemical components from known herbs and extracts. Their aim was to identify a single chemical that was responsible for the extract's pharmacological effects—the active principle. For example, morphine is the active principle responsible for the sedative properties of opium, while cocaine is the active principle present in coca leaves. The active principle in willow bark is salicyclic acid. Other active principles isolated in the 19th century included quinine, caffeine, atropine, physostigmine, and theophylline. Quinine was of particular importance since it is effective in treating malaria. Caffeine and theophylline are stimulants found in beverages. Atropine has been used in cardiovascular medicine and as an antidote to pesticide poisoning, while physostigmine can be used to treat glaucoma.

It was not long before chemists synthesized analogues of active principles. Analogues are structures which have been modified slightly from the original active principle. Such modifications can often improve activity or reduce side effects. This led to the concept of the lead compound—a compound with a useful pharmacological activity that could act as the starting point for further research. Fully synthetic compounds were also investigated for pharmacological activity leading to the discovery of general anaesthetics, local anaesthetics, and barbiturates during the late 19th and early 20th centuries.

The first half of the 20th century culminated in the discovery of effective antimicrobial agents. At the beginning of the century, Paul Ehrlich developed arsenic-containing drugs which proved effective against syphilis, while early antimalarial agents were discovered in the 1920s. The sulphonamides were discovered in the 1930s, but the most important advance was penicillin, which was introduced in the 1940s. The original penicillin was isolated from a fungus, and this sparked a massive worldwide study of fungal cultures in the post-war years, which led to the identification of many of the antibiotics used in medicine today. The middle part of the 20th century was a golden age of antibacterial research, and marked one of the most important advances in medicine. Before the antibiotic revolution, even simple wounds could prove life-threatening, and many of the surgical operations carried out routinely today were totally impractical.

The development of rational drug design

The 1960s can be viewed as the birth of rational drug design. During that period there were important advances in the design of effective anti-ulcer agents, anti-asthmatics, and beta-blockers for the treatment of high blood pressure. Much of this was based on trying to understand how drugs work at the molecular level and proposing theories about why some compounds were active and some were not.

However, rational drug design was boosted enormously towards the end of the century by advances in both biology and chemistry. The sequencing of the human genome led to the identification of previously unknown proteins that could serve as potential drug targets. For example, kinase enzymes have proved important targets for novel anti-cancer agents in recent years. These enzymes catalyse phosphorylation reactions and play a key role in controlling cell growth and division. Similarly, the sequencing of genomes from viruses led to the identification of viral-specific proteins that could serve as novel targets for new antiviral agents. Advances in automated, small-scale testing procedures (high-throughput screening) also allowed the rapid testing of potential drugs.

In chemistry, advances were made in X-ray crystallography and NMR spectroscopy, allowing scientists to study the structure of drugs and their mechanisms of action. Powerful molecular modelling software packages were developed that allowed researchers to study how a drug binds to a protein binding site. Novel synthetic methods have boosted the ability of chemists to create new compounds. In addition, the development of automated synthetic methods has vastly increased the number of compounds that can be synthesized in a given time period. Companies can now produce thousands of compounds that can be stored and tested for pharmacological activity. Such stores have been called chemical libraries and are routinely tested to identify compounds capable of binding with a specific protein target. These advances have boosted medicinal chemistry research over the last twenty years in virtually every area of medicine.

There has also been a significant change in the way pharmaceutical research is tackled. For most of the 20th century, drug research depended on the discovery of a lead compound with a useful pharmacological activity. Thousands of analogues were then synthesized in an effort to find an improved compound. Years later, the molecular target might be discovered, allowing a better understanding of the biological mechanisms affected by these agents. In this approach, progress was dictated by whatever lead compound was discovered.

Nowadays, most research projects are initiated by choosing a potential drug target, such as an enzyme or a receptor. A lead compound that interacts with that protein target is then sought. Of course, there are still research projects that are determined by the chance discovery of a pharmacologically active compound, but the scientific approach towards the design of novel drugs now follows the pathway shown in Figure 54. Those stages requiring knowledge of organic chemistry are highlighted in bold.

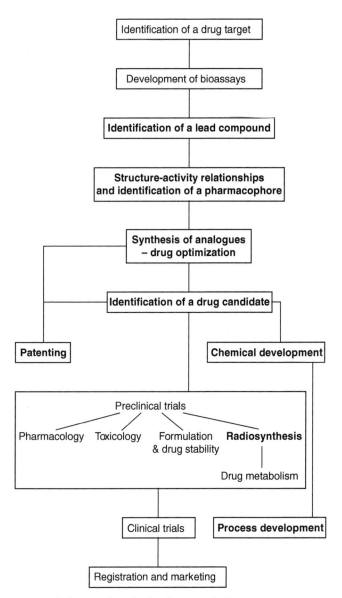

54. **A typical approach to the development of a drug.**

Identification of a drug target

Drugs interact with molecular targets in the body such as proteins and nucleic acids. However, the vast majority of clinically useful drugs interact with proteins, especially receptors, enzymes, and transport proteins (Chapter 4).

Drugs can be designed to activate receptors in the same way as the natural messenger. Such drugs are called agonists. Alternatively, drugs can be designed to block a receptor without activating it. Such drugs are known as antagonists. Examples of receptor antagonists include the beta-blocker propranolol, and the anti-ulcer agents cimetidine and ranitidine. Examples of receptor agonists include the anti-asthmatic drug salbutamol and the analgesic morphine.

Enzymes are also important drug targets. Drugs that bind to the active site and prevent the enzyme acting as a catalyst are known as enzyme inhibitors. Examples of enzyme inhibitors include the anti-HIV drug saquinavir, and the anti-hypertensive agent captopril. Enzymes are located inside cells, and so enzyme inhibitors have to cross cell membranes in order to reach them—an important consideration in drug design. There is no point designing a potent enzyme inhibitor if it fails to cross the cell membrane.

Transport proteins are targets for a number of therapeutically important drugs. For example, a group of antidepressants known as selective serotonin reuptake inhibitors prevent serotonin being transported into neurons by transport proteins. As a result, serotonin levels increase and produce the observed antidepressant effect.

Drug testing and bioassays

Bioassays are tests that are used to identify whether a drug interacts with a protein target. *In vitro* tests are carried out on target molecules or cell cultures. For example, enzyme inhibitors can be tested on a purified enzyme in a test tube to see whether they prevent the enzyme catalysing a particular reaction. *In vitro* tests can be automated, allowing the rapid testing of thousands of compounds in a very small time period. This is known as high-throughput screening. These tests are ideal for identifying whether drugs interact with a molecular target to produce a particular pharmacological effect. The ability of a drug to bind to its target and produce such an effect is known as pharmacodynamics.

In vivo bioassays are carried out on living organisms, and are complimentary to *in vitro* tests. *In vivo* tests establish whether a drug produces a physiological effect, such as analgesia or the lowering of blood pressure. *In vivo* tests also establish whether a drug reaches its molecular target when it is administered to an organism. The range of factors affecting a drug's ability to reach its target is known as pharmacokinetics.

The main pharmacokinetic factors are absorption, distribution, metabolism, and excretion. Absorption relates to how much of an orally administered drug survives the digestive enzymes and crosses the gut wall to reach the bloodstream. Once there, the drug is carried to the liver where a certain percentage of it is metabolized by metabolic enzymes. This is known as the first-pass effect. The 'survivors' are then distributed round the body by the blood supply, but this is an uneven process. The tissues and organs with the richest supply of blood vessels receive the greatest proportion of the drug. Some drugs may get 'trapped' or sidetracked. For example fatty drugs tend to get absorbed in fat tissue and fail to reach their target. The kidneys are chiefly responsible for the excretion of drugs and their metabolites. They are particularly efficient at excreting polar molecules.

In vivo tests can sometimes identify unexpected activity that would not be picked up by *in vitro* tests. For example, the dye prontosil was shown to have antibacterial activity *in vivo*, but proved inactive *in vitro*. This is because prontosil itself is inactive, and is metabolized to an active sulphonamide within the body. A compound that acts in this way is known as a prodrug.

Finally, *in vivo* tests can detect side effects which would not be observed by *in vitro* tests. Sometimes, this suggests unexpected applications for the drug. For example, the anti-impotence drug sildenafil was originally tested as an anti-hypertensive drug, and its anti-impotence effects were only identified during early clinical trials.

Identification of lead compounds

A lead compound is a chemical structure that can bind to a desired molecular target. It may not bind particularly strongly and it may not be particularly active, but the fact that it binds to the desired target means that it can serve as a starting point for further research. The medicinal chemist can then 'tweak' the structure to find analogues that bind more strongly and have better activity and selectivity.

Lead compounds are obtained from both the natural world and the laboratory. Historically, the natural world has been a rich source of novel lead compounds and remains so today. However, finding them is usually a slow process. Moreover, there is no guarantee of success. Nowadays, there is much greater emphasis on generating lead compounds by synthesis or rational design.

Structure–activity relationships and pharmacophores

Having identified a lead compound, it is important to establish which features of the compound are important for activity. This, in turn, can give a better understanding of how the compound binds to its molecular target.

Most drugs are significantly smaller than molecular targets such as proteins. This means that the drug binds to quite a small region of the protein—a region known as the binding site (Figure 49). Within this binding site, there are binding regions that can form different types of intermolecular interactions such as van der Waals interactions, hydrogen bonds, and ionic interactions. If a drug has functional groups and substituents capable of interacting with those binding regions, then binding can take place.

A lead compound may have several groups that are capable of forming intermolecular interactions, but not all of them are necessarily needed. One way of identifying the important binding groups is to crystallize the target protein with the drug bound to the binding site. X-ray crystallography then produces a picture of the complex which allows identification of binding interactions. However, it is not always possible to crystallize target proteins and so a different approach is needed. This involves synthesizing analogues of the lead compound where groups are modified or removed. Comparing the activity of each analogue with the lead compound can then determine whether a particular group is important or not. This is known as an SAR study, where SAR stands for structure–activity relationships.

Once the important binding groups have been identified, the pharmacophore for the lead compound can be defined. This specifies the important binding groups and their relative position in the molecule. The pharmacophore can be represented by highlighting the binding groups on the lead compound. For example, the pharmacophore for estradiol consists of three functional groups—the phenol group, the aromatic ring, and the alcohol group (Figure 55). The remainder of the tetracyclic structure serves as a rigid scaffold to hold the important binding groups in the correct positions, such that they interact simultaneously with the target binding site.

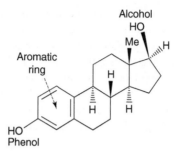

55. The pharmacophore for estradiol.

Identifying the pharmacophore of a rigid molecule such as estradiol is relatively straightforward, but with flexible compounds it is difficult to determine the relative position of important binding groups since the drug can exist in different shapes (conformations). For example, the binding groups for dopamine (an important neurotransmitter in the brain) are the two phenol groups, the aromatic ring, and the charged amine group (Figure 56). The relative positions of the phenol groups and the aromatic ring are easy to define as this is a rigid part of the molecule, but the position of the charged amine cannot be

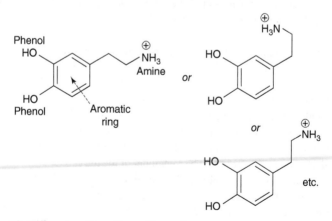

56. Different conformations of dopamine.

defined. This is because the bonds in the side chain can rotate, producing a large number of possible conformations. The conformation that binds most effectively to the binding site will have the charged amine group positioned in a particular way, and this is known as the active conformation. Other conformations will bind less effectively.

One way of identifying the active conformation of a flexible lead compound is to synthesize rigid analogues where the binding groups are locked into defined positions. This is known as rigidification or conformational restriction. The pharmacophore will then be represented by the most active analogue. For example, the structures shown in Figure 57 are rigid analogues of dopamine which have particular conformations of dopamine trapped within their structure. The bold bonds highlight the three-carbon chain that is present in dopamine. If one of these analogues proved significantly more active than the others, it can be used to define the active conformation and the pharmacophore.

A large number of rotatable bonds is likely to have an adverse effect on drug activity. This is because a flexible molecule can adopt a large number of conformations, and only one of these shapes corresponds to the active conformation. If the molecule enters the binding site in an inactive conformation, it will depart again without binding. Indeed, a flexible molecule may have to enter and depart a binding site several times before it adopts the correct active conformation for binding. In contrast, a totally rigid molecule

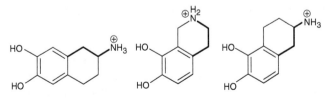

57. Rigid analogues of dopamine.

containing the required pharmacophore will bind the first time it enters the binding site, resulting in greater activity.

Drug design and drug optimization

Drug optimization involves designing and synthesizing analogues of the lead compound in order to find structures with improved activity, selectivity, and pharmacokinetics. A crystal structure of the lead compound bound to the target protein will help enormously in this quest—a process known as structure-based drug design—but it is not always possible to crystallize the target protein. Fortunately, there are a number of well-established design strategies that can aid the medicinal chemist in deciding which analogues are worth synthesizing. The strategy of designing a rigid analogue that mimics the active conformation of the lead compound has already been mentioned. Another strategy is to add extra groups to the structure, which would allow additional binding interactions to take place with parts of the binding site not occupied by the lead compound.

It is also important to optimize a drug's pharmacokinetic properties such that it can reach its target in the body. Strategies include altering the drug's hydrophilic/hydrophobic properties to improve absorption, and the addition of substituents that block metabolism at specific parts of the molecule.

Drug candidates and patenting

The drug optimization process produces a large number of compounds, several of which could be considered as a drug candidate for preclinical tests and clinical trials. Several factors are involved in deciding which one goes forward. The drug candidate must have useful activity and selectivity, with minimal side effects. It must have good pharmacokinetic properties, lack toxicity, and preferably have no interactions with other drugs that might be taken by a patient. Finally, it is important that it can be synthesized as cheaply as possible in order to make the maximum profit. Therefore, if there is a choice between two compounds of similar activity, then the choice may well be determined by identifying which one is cheaper to synthesize.

Once a promising-looking drug has been identified, it needs to be patented such that the company has exclusive rights to marketing. Since patenting takes place relatively early on in the drug development process, several years of the patent are lost due to the time taken to carry out preclinical and clinical trials.

Patenting drugs can lead to ethical dilemmas since the majority of people in the developing world cannot afford them. To address this, the World Trade Organization's TRIPS (trade-related aspects of intellectual property rights) agreement allows governments in the developing world to grant compulsory licences for the manufacture of potentially life-saving drugs. This allows a country to bypass patent regulations and produce urgently needed medicines for its own citizens. Unfortunately, some countries have stretched the definition of life-threatening conditions. For example, in 2012, India imposed a compulsory licence on sorafenib—an anti-cancer agent which is considered to be life extending rather than life saving. This has created concerns that pharmaceutical companies might stop developing pharmaceuticals in therapeutic areas such as cancer or tropical diseases.

Chemical and process development

Once a candidate drug has been identified, work starts on developing a large-scale synthesis that will provide sufficient quantities of the drug for preclinical and clinical trials. This is known as chemical development. The development chemist has a demanding role as it is necessary to produce large quantities of the drug as quickly as possible, while maintaining the quality of each batch produced. This is crucial since the preclinical tests and clinical trials must be carried out on batches having a consistent purity.

Otherwise, the tests are not comparing like with like. The chemical development process is more than just scaling up the original synthesis. Reactions may need to be modified, or altered altogether, in order to optimize yields. Indeed, the final production synthesis may be totally different from the original research synthesis.

Preclinical trials and formulation

Preclinical trials involve testing the drug candidate for selectivity, toxicity, and possible side effects. Most of this work is carried out by toxicologists, pharmacologists, and biochemists. However, organic chemists are needed to synthesize samples of the drug containing a radioisotope such as ^{14}C. Such radiolabelled compounds are used to monitor the distribution and metabolism of the drug during *in vivo* tests.

Formulation involves pharmacists and pharmaceutical chemists who identify how best to store and administer the drug, for example as a pill or capsule.

Clinical trials and regulatory affairs

Clinical trials are the province of the clinician. There are four phases of clinical trials. Phase 1 involves a small group of healthy volunteers, whereas the later phases involve patients. The clinical trials are the most expensive and time-consuming part of the process in getting a drug to market and many drugs fail the process. This can be because they are not sufficiently effective, or they produce unacceptable side effects.

Regulatory bodies such as the US Food and Drug Administration (FDA) and the European Agency for the Evaluation of Medicinal Products monitor the process, and have to give their approval before the drug is finally marketed.

The future

Since the 1980s, there has been significant progress in treating diseases that were once considered untreatable. For example, there has been remarkable progress in designing effective antiviral drugs, inspired by the need to treat HIV. There has also been significant progress in treating various cancers. The development of a class of drugs called kinase inhibitors has been particularly important in this respect. However, there are some diseases that are still proving to be problematic. There are no cures for Alzheimer's or Parkinson's disease, and finding treatments for these diseases provides major challenges for the future. Most drugs that have reached clinical trials for the treatment of Alzheimer's disease have failed. Between 2002 and 2012, 244 novel compounds were tested in 414 clinical trials, but only one drug gained approval. This represents a failure rate of 99.6 per cent as against a failure rate of 81 per cent for anti-cancer drugs.

The increase in drug-resistant bacterial strains is another concern. Several bacterial strains (such as *Staphylococcus aureus*) gain resistance to antibacterial agents due to their relatively high rates of mutation. For example, in the 1960s, *S. aureus* strains emerged that had gained resistance to the early penicillins. This crisis was averted by designing a new penicillin called methicillin that could combat these strains, but further strains have now developed that are resistant to methicillin (MRSA, or methicillin-resistant *S. aureus*). Other problem infections include multidrug-resistant tuberculosis and *E. faecalis*. Therefore, it is important to continue the search for novel antibacterial agents.

Several approaches to finding new antibacterial agents are being investigated. For example, the pharmaceutical company GlaxoSmithKline is currently investigating compounds that inhibit a bacterial enzyme called polypeptide deformylase. It has also been suggested that research should move away from finding new broad-spectrum antibacterial drugs that treat a wide variety of infections, to designing drugs that target specific infections. This is because broad-spectrum drugs have proved more susceptible to drug resistance. A therapy that combines a number of 'narrow-spectrum' drugs that act on different targets could be effective because there is little chance of drug resistance arising to all the drugs present in the combination.

Unfortunately, the trend in recent decades has been to cut back on antibacterial research due to the relatively low rates of success in obtaining new agents. Moreover, any novel agents that *are* discovered are likely to be placed on a reserve list to limit the chances of resistance arising. Consequently, pharmaceutical industries are unlikely to gain a significant financial return for the huge research investment required to design new drugs. Various national and world organizations have recognized the dangers and are now warning that more research is needed. In April 2014, the World Health Organisation declared that urgent, coordinated action was needed on a global scale to stop the world entering a post-antibiotic era where bacterial infections could once again become untreatable and result in even the most minor of injuries being potentially fatal.

Antimicrobial resistance is now seen as a ticking time bomb of global proportions, which poses as serious a threat to civilization as climate change. Tackling that threat is likely to require coordinated and funded collaborations between governments, pharmaceutical companies, and academic institutions. Fortunately, governments have now recognized the threat and have introduced new initiatives to encourage collaborative research.

Veterinary drugs

The same strategies used to design human medicines are applied to the development of veterinary drugs. Veterinary drugs are often different from human drugs because animals have different biochemical and metabolic systems. A compound that is safe for humans may be toxic to an animal. For example, theobromine is a constituent of chocolate that is toxic to dogs, but not to humans. 'Doggy chocolates' have to be specially formulated to lack theobromine.

Different animal species may also require different medicines for a specific disease. A further complication involves medicines for farm animals, since traces of the medication could end up in the food we eat. Therefore, there are regulations determining how long farmers should wait before slaughtering or milking animals that have received drug treatment. The discovery of horse meat in some European meat products during 2013 was a concern because it raised the possibility that traces of veterinary drugs might be present in the meat. For example, phenylbutazone is an anti-inflammatory agent used on horses, but causes adverse effects in humans.

The use of antibacterial agents in veterinary medicine is another concern since this could increase the prevalence of antibacterial resistance. For that reason, it is best to use agents that are not used in human medicine. Controversially, antibiotics such as penicillins and cephalosporins have been used to promote animal growth, but legislation is being passed in various countries to prevent this practice.

Each drug that is used in veterinary practice is approved for specific species, which means that a drug approved for use on a dog is not necessarily approved for use on a cat. To date, there are 634 drugs approved for dogs and 313 for cats. A complication in the treatment of dogs relates to different breeds. Some breeds are more susceptible to certain diseases than others, and there can be differences in drug metabolism. For example, the antiparasitic drug ivermectin is licensed for dogs, but can prove toxic to collies. As far as livestock is concerned, there are 688 products available for cattle in the USA, most of which are used to treat infections or inflammation. Veterinary medicine even includes medicines for bees. The varroa mite is one of the factors thought to be responsible for colony collapse disorder, where worker bees suddenly disappear from the hive. The mite can be treated with pyrethroid and organophosphate pesticides (Chapter 6), but mild cases of infection can be treated with antibiotics such as oxytetracycline.

Drugs of abuse

Several drugs that were once lauded as medical breakthroughs are now classed as drugs of abuse. For example, heroin was marketed at the end of the 19th century and was hailed as the 'heroic' drug that would eliminate pain. Unfortunately, nobody anticipated its addictive properties. Similarly, Sigmund Freud advocated the use of cocaine as an antidepressant until its addictive properties became clear. The psychoactive drug LSD was originally introduced as a medicine, and, in the 1970s, a psychoactive compound called MDMA was investigated as an aid to psychotherapy. However, the drug's euphoric effects means that it is now taken as a 'social drug'. This is the drug now known as 'ecstasy'.

These are examples of drugs that were originally developed with the best of intentions, but a number of unscrupulous chemical companies are now deliberately designing drugs of abuse. These include stimulants that act in the same way as amphetamines. Because they are novel structures, they are not illegal and it is legitimate to sell them, as long as they are not advertised for human consumption. Instead, they are advertised as bath salts, plant food, or window cleaner.

Such designer drugs have been labelled 'legal highs', which might mislead consumers into thinking that they are legally approved. However, none of these drugs have undergone the preclinical and clinical trials required for medicines. Anyone taking them is gambling with their health, if not their life. When the UK government banned the legal high 'serotoni', it had already been responsible for thirty-seven deaths in the UK alone. Another forty-two people died taking the stimulant mephedrone.

Unfortunately, it takes time to identify novel legal highs, and even longer to make them illegal. By the time a 'legal high' has been banned, the company that produces it has usually modified the structure and introduced a new stimulant. For example, the product called Ivory Wave was sold as bath salts, and contained a psychoactive compound called methylenedioxyprovalerone. When this compound was banned, it was replaced with a similar structure called naphthylpyrovalerone. When this was banned, desoxypipradol was added instead. Desoxypipradol is more potent than the previous two compounds, and many regular users of Ivory Wave overdosed on the product.

The problem has been increasing in recent years. In 2009, there were twenty-four 'legal highs' sold across Europe, but this had risen to eighty-one by 2013. Chinese labs are thought to be producing most of the legal highs that are currently available. There has also been a rise in synthetic cannabinoids such as 'annihilation' which was responsible for nine people being hospitalized in 2012.

The UK government has now passed a law that bans any substance capable of producing a psychoactive effect, rather than banning each structure as it appears. However, this may only serve to drive the market underground. Moreover, genuine research on psychoactive drugs may be hindered by the need for government licences. There could be even wider ramifications if chemical suppliers feel obliged to stop selling chemicals used in the synthesis of legal highs, as that would adversely affect many legitimate research projects.

Chapter 6
Pesticides

Pesticides are organic chemicals produced by the agrochemical industry to improve agricultural yields and to fight crop disease. They include insecticides, fungicides, and herbicides, which have proved vital in increasing food production for a global population that is expected to increase by 33 per cent over the next thirty-five years. Without pesticides, food production could only be maintained by increasing the amount of land committed to crops, but this would mean turning over much of the world's prairies, woodlands, grasslands, and meadows to agriculture—a strategy that would adversely affect biodiversity and cause unpredictable effects on the ecosystem.

Concerns over the effects of traditional pesticides has promoted research into designing safer and more environmentally friendly pesticides—a role that falls to organic chemists. It takes about ten years and £160 million to develop a new pesticide, and only large companies can afford that level of investment. Several companies specialize in agrochemical production and research, and there is a vast global market for agrochemicals. The volume of global sales increased 47 per cent in the ten-year period between 2002 and 2012, while, in 2012, total sales amounted to £31 billion. The biggest markets were in Brazil, the USA, and Japan.

In many respects, agrochemical research is similar to pharmaceutical research. The aim is to find pesticides that are toxic to 'pests', but relatively harmless to humans and beneficial life forms. The strategies used to achieve this goal are also similar. Selectivity can be achieved by designing agents that interact with molecular targets that are present in pests, but not other species. Another approach is to take advantage of any metabolic reactions that are unique to pests. An inactive prodrug could then be designed that is metabolized to a toxic compound in the pest, but remains harmless in other species. Finally, it might be possible to take advantage of pharmacokinetic differences between pests and other species, such that a pesticide reaches its target more easily in the pest.

Insecticides

Prior to World War II, only naturally occurring insecticides were available. For example, sulphur was used for pest control in ancient Greece, and is still used in some parts of the world today. In 1690, it was reported that tobacco extracts were effective in controlling insects, and in the early 1800s, other plant extracts were used for their insecticidal properties—namely pyrethrins from chrysanthemums and rotenone from derris roots. More recently, extracts from an Indian plant called the neem tree have proved effective. The agents responsible for the insecticidal activity of these extracts have been identified as nicotine from tobacco plants, pyrethrum from chrysanthemums, and azadirachtin from the neem tree.

Natural products are limited in their availability, selectivity, and effectiveness, and so it was not until the advent of synthetic insecticides that potent, selective, and affordable insecticides became available on an industrial scale. The early synthetic insecticides included organochlorines, organophosphates, methylcarbamates, and pyrethroids. In general, these agents were potent and showed selective toxicity against insects rather than mammals. However, their cumulative effects on the environment and other life forms were not fully anticipated at the time. They have now been largely replaced by insecticides that are more selective and environmentally friendly.

Insecticides: organochlorine agents

Organochlorine agents were the first synthetic insecticides to reach the market starting with DDT (Figure 58). DDT was first synthesized in 1874. However, its properties as an insecticide were not discovered until 1939, when it was found to be effective against mosquitoes, ticks, and locusts. This led to it being used by the military during World War II to combat malaria in South-East Asia, and typhus in Eastern Europe. After the war, DDT played a major role in eliminating malaria from Europe and North America, and, at one point, there were hopes that it might even eradicate malaria worldwide. Unfortunately, resistance to the chemical developed in more tropical areas.

Nevertheless, DDT has saved a vast number of lives from insect-borne diseases such as malaria, yellow fever, and sleeping sickness. One estimate puts the number of lives saved as 500 million between the 1940s and 1960s. As a result, the 1948 Nobel Prize in Medicine was awarded to Paul Muller—the man who discovered DDT's insecticidal properties.

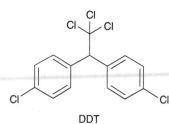
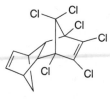

DDT

Aldrin

58. Examples of organochlorine insecticides.

As well as combating disease, DDT found widespread use as an agricultural insecticide, with an average of 40,000 tons being produced each year. DDT is very toxic to insects, but has a much lower toxicity to mammals. In fact, the lethal dose for a human would be equivalent to the amount of DDT required to treat an acre of land. Unfortunately, despite the undoubted benefits of DDT in terms of lives saved and increased crop production, there was a heavy environmental cost. DDT is a relatively stable molecule, and so it accumulates in the environment. It is also hydrophobic in nature, which means that it is poorly soluble in water, but dissolves easily in the body fat of various forms of wildlife. Later research revealed that the concentration of DDT in wildlife increased the higher one goes up the animal food chain, and this proved particularly devastating for predatory birdlife. DDT was blamed for the near extinction of the bald eagle and the peregrine falcon in the USA, since it was found to cause egg shell thinning. These fragile eggs tended to break before hatching occurred, killing the embryos.

DDT was banned from agricultural use in the USA in 1972, with the UK following suit in 1984. A worldwide ban was introduced in 2004 by the Stockholm Convention, but DDT is still permitted for the eradication of insects if they are likely to be harmful to human health. For example, DDT is still used in India to control malaria.

Another example of an organochloride insecticide is aldrin (Figure 58). Like DDT, aldrin contains several chlorine atoms and is a hydrophobic molecule, but it has a totally different carbon skeleton in the form of a complex multi-ring system.

The organochlorine insecticides act on ion channels, resulting in the disruption of nerve transmission, spasms, and death. DDT acts on sodium ion channels, whereas aldrin and its analogues act on chloride ion channels. Because they act on different types of ion channels, resistance to DDT does not result in cross-resistance to aldrin. Resistance develops when mutations alter amino acids in the target ion channels. This, in turn, weakens binding interactions with insecticides.

Insecticides: methylcarbamates and organophosphates

The methylcarbamates and organophosphates were developed after the organochlorides. The design of methylcarbamates was based on a natural product called physostigmine (Figure 59), which is a poison found in calabar beans. Physostigmine inhibits an enzyme called acetylcholinesterase, which catalyses the hydrolysis of a neurotransmitter called acetylcholine (Figure 60). When the enzyme is inhibited, acetylcholine levels increase and over-stimulate protein receptors in the insect's nervous system, leading to toxicity and death. The acetylcholinesterase enzyme is also present in humans, and so it is crucial that the methylcarbamate insecticides are selectively toxic to the insect version of the enzyme. Physostigmine itself does not have this selectivity, but carbaryl is an analogue that does.

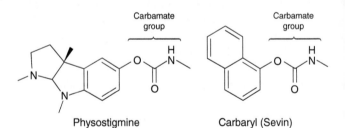

59. Physostigmine and the insecticide carbaryl.

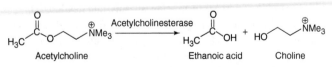

60. Hydrolysis of acetylcholine catalysed by acetylcholinesterase.

The organophosphates also target the acetylcholinesterase enzyme, and act as irreversible inhibitors. Several organophosphates are too toxic to be used as insecticides. Indeed, several have been used as nerve toxins in chemical warfare. These include sarin, tabun, soman, dyflos, and VX. In all these examples, the poison occupies the active site of the enzyme, then reacts with a serine residue. A phosphate group is transferred from the nerve agent to a serine residue and 'caps' it such that it can no longer catalyse the hydrolysis of acetylcholine (Figure 61).

Considering the toxicity of nerve agents, it seems a tall order to design a safe organophosphate insecticide. In fact, this was achieved by designing prodrugs that were only metabolized in insects to the active compound. Parathion, malathion, and chlorpyrifos (Figure 62) are insecticides that contain a P=S group. As such, they have no direct effect on the acetylcholinesterase enzyme. However, insects have a metabolic enzyme that alters the P=S group to a P=O group. The resulting nerve agent can then inhibit acetylcholinesterase.

In mammals, these insecticides are metabolized by different enzymes to give inactive compounds which are excreted. Despite this, organophosphate insecticides are not totally safe, and

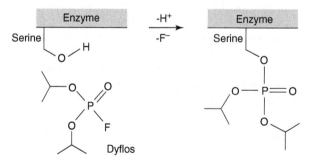

61. Reaction of dyflos with a serine residue in the active site of acetylcholinesterase.

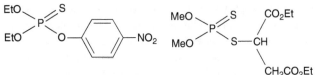

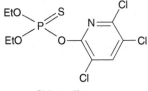

Chlorpyrifos

62. Organophosphate prodrugs used as insecticides.

prolonged exposure can cause serious side effects if they are not handled with care. They also have a cumulative toxic effect on wildlife, and so alternative agents are now favoured.

Insecticides: pyrethrins and pyrethroids

Pyrethrum is a plant extract obtained by crushing chrysanthemums in water, and contains a mixture of natural products called pyrethrins. Such extracts have been used as insecticides and insect repellents for many years, and it is believed that the Chinese may have used them as early as 1000 BC. French soldiers reportedly used chrysanthemums to repel fleas and lice during the Napoleonic wars. Six pyrethrins with very similar structures have been identified so far (Figure 63). Like DDT, they bind to sodium ion channels in an insect's nervous system, resulting in paralysis and death. A potential problem with pyrethrins is the fact that they act on the same target as DDT. This means that any pests that acquire resistance to DDT are often resistant to pyrethrins—an example of cross-resistance.

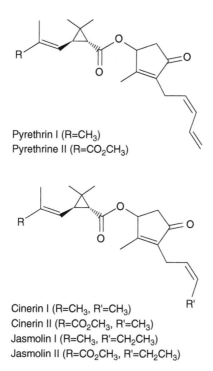

Pyrethrin I (R=CH₃)
Pyrethrine II (R=CO₂CH₃)

Cinerin I (R=CH₃, R'=CH₃)
Cinerin II (R=CO₂CH₃, R'=CH₃)
Jasmolin I (R=CH₃, R'=CH₂CH₃)
Jasmolin II (R=CO₂CH₃, R'=CH₂CH₃)

63. Structures of pyrethrins.

Combining pyrethrins with the synthetic additives piperonyl butoxide or sesamex (Figure 64) allows pyrethrins to be effective against a wider range of insects, including those that are normally resistant. This is because the synthetic additives inhibit those enzymes in the insect that normally metabolize and deactivate the pyrethrins. An agent that enhances the activity of another is called a synergist. One disadvantage with a synergist is that it could potentially inhibit mammalian metabolic enzymes and increase susceptibility to toxins.

The pyrethrins are considered to be among the safest insecticides on the market. Consequently, several household pesticides contain

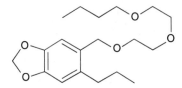

Piperonyl butoxide

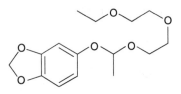

Sesamex

64. Examples of synergists.

them. They are also biodegradable when exposed to light or oxygen (unlike DDT), and form harmless products. Unfortunately, pyrethrins are harmful to bees, and so they should be applied at night when bees are not pollinating.

The pyrethroids are synthetic analogues of pyrethrins and were introduced to replace organochlorines during the 1950s. They are not as biodegradable as pyrethrins, which makes them more effective as insecticides, but this also makes them more likely to accumulate in the environment. Some commercial insecticides and shampoos contain both pyrethrins and pyrethroids, and there is an element of risk in their use, especially in terms of allergies. Examples of synthetic pyrethroids include phenothrin and cypermethrin (Figure 65).

Insecticides: neonicotinoids

Nicotine (Figure 66) has insecticidal properties because it activates a type of cholinergic receptor called the nicotinic receptor. As a

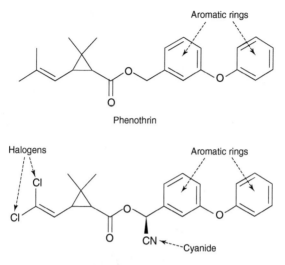

Aromatic rings

Phenothrin

Halogens

Aromatic rings

CN ◄----Cyanide

Cypermethrin (1978)

65. Examples of pyrethroids.

result, nerves are overstimulated, resulting in toxic effects. Although nicotine has been used as an insecticide in the form of tobacco extracts, it is not as potent as synthetic insecticides, and shows poor selectivity between the cholinergic receptors in insects and those in mammals. A large range of structurally related analogues were synthesized to try and find compounds with improved selectivity, but without success. It was not until a structurally unrelated lead compound was discovered that potent and selective agents could be developed. These also bind to the nicotinic receptor, and have been dubbed neonicotinoids.

The development of the neonicotinoids began in 1970, and led ultimately to the development of imidacloprid (Figure 66) which proved 10,000 times more active than nicotine. It was patented in 1985 and introduced to the market in 1991. The compound proved

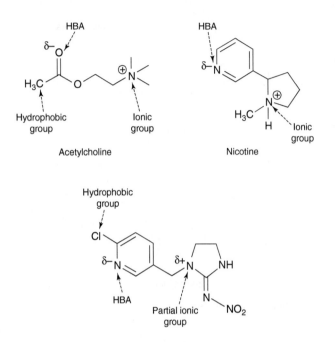

66. Important binding groups in acetylcholine, nicotine, and imidacloprid. HBA stands for hydrogen bond acceptor (see Chapter 2).

extremely successful and it was the first insecticide to reach sales of a billion dollars per year. It soon became the most widely used insecticide in the world, and its development has been described as a milestone in insecticidal research. As well as its use as an agricultural insecticide, it is used in veterinary practice to control ticks and fleas. Several other neonicotinoids have since been developed and they have become established as the most important class of insecticides on the market.

Imidacloprid, nicotine, and acetylcholine share structural features that are important in binding these agents to the receptor binding site (Figure 66). They all contain a positively charged or partially positively charged ($\delta+$) nitrogen that can interact with the receptor binding site through ionic interactions. In addition, they all contain an atom with a slight negative charge ($\delta-$), which can form a hydrogen-bonding interaction with the binding site. Imidacloprid can form additional binding interactions because the hydrophobic chlorine substituent fits into a hydrophobic pocket in the binding site.

However, this does not explain why imidacloprid binds 1,000 times more strongly to insect receptors than human receptors—a key reason for its selectivity. A major reason for this selectivity is the presence of the nitro group, which can interact with an arginine residue that is present in the binding site of insect receptors but not mammalian receptors. A second reason is the lack of a fully positively charged nitrogen atom, which weakens the ionic interactions with the mammalian receptor. Pharmacokinetic factors also play a role in enhancing selectivity. Imidacloprid can cross the blood brain barrier of insects to attack their central nervous system, but it is unable to cross the blood brain barrier of mammals.

Neonicotinoids were originally thought to have low toxicity to bees, but many now feel that they have been responsible for a rapid decline in honey bee numbers since 2006—a phenomenon called colony collapse disorder. The situation became so serious that commercial bee-keepers in the USA had lost up to half their hives by 2012. The impact on agriculture was even greater since it is estimated that bees are responsible for pollinating US crops worth £9.8 billion. It has been suggested that neonicotinoids may affect the ability of bees to forage, learn, and remember navigation routes to and from food sources.

In 2013, the European Union decided to restrict the use of neonicotinoids until December 2015, and recommended that they be restricted to crops that did not attract bees. The US followed suit soon afterwards. This certainly marked a victory for environmentalists, but several scientists claim that the decision was taken as much for political reasons as scientific. The company Bayer Cropscience, which produces two of the three restricted products, stated that neonicotinoids are safe for bees, as long as they are used responsibly. They also argued that the decline in bee populations had begun before neonicotinoids were introduced and could be due to many other factors such as mites that carry viruses, fungal disease, and a reduction in flowering plants due to increasing agriculture. They also highlighted the fact that Australia has very healthy bees despite the widespread use of neonicotinoids. This could well be due to the lack of the varroa mite in Australia compared to its prevalence in Europe. In truth, it is difficult to be sure which specific factor is most important in causing the decline in bee populations, and it is perfectly possible that a combination of several factors are important.

Banning neonicotinoids brings its own risks. Farmers might be forced to revert back to older forms of insecticide which are more environmentally damaging. This could also increase the emergence of resistance towards these agents. Regardless of who is right or wrong, the demands for safer and more selective insecticides continue to challenge the ingenuity of research chemists. As a result, other insecticides have been developed or are under investigation. For example, a number of different structures act on the nicotinic cholinergic receptor. One of these involves a group of agents that are activated by insect metabolism to form nereistoxin—a naturally occurring neurotoxin produced by a marine annelid worm. Other compounds include the spinosyns, which are complex natural products produced by a bacterial strain called *Saccharopolyspora spinosa*, sulfoximines, and analogues of a natural product called stemofoline. One such structure is flupyradifurone, which has recently been approved (Figure 67).

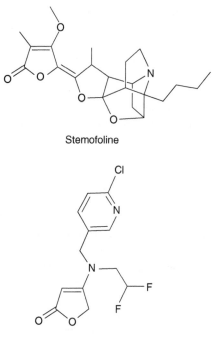

Stemofoline

Flupyradifurone

67. Stemofoline and flupyradifurone.

Future insecticides

Insecticides are being developed that act on a range of different
targets as a means of tackling resistance. If resistance should arise
to an insecticide acting on one particular target, then one can
switch to using an insecticide that acts on a different target.
Ryanoids such as chlorantraniliprole, cyantraniliprole, and
flubendiamide were marketed in 2006–7. They act by binding to
calcium ion channels in muscle cells and cause paralysis and
death. They show high selectivity for insects over mammals, and
have very little toxicity towards birds or aquatic organisms. More

recent research at the University of Newcastle has discovered a naturally occurring peptide that also targets calcium ion channels. The peptide is found in the venom of the Australian funnel web spider, and is toxic to aphids and caterpillars, but harmless to mammals and honey bees.

Several insecticides act as insect growth regulators (IGRs), and target the moulting process rather than the nervous system. In general, IGRs take longer to kill insects but are thought to cause less detrimental effects to beneficial insects. A juvenile insect grows a new exoskeleton under its old one, then sheds the old exoskeleton by moulting. This allows the new exoskeleton to swell and harden.

There are two major hormones involved in the moulting process—juvenile hormone and ecdysone. Eight varieties of juvenile hormone have been identified from different insect species, but they all contain a methyl ester at one end of the chain and an epoxide at the other end (Figure 68). Juvenile hormones must be absent if a pupa is to moult into an adult, and so several IGRs prevent insects maturing into adults by mimicking juvenile hormones. The IGRs used to mimic juvenile hormone are called juvenoids and are structural analogues of the juvenile hormone. Methoprene is the most successful of these and is considered sufficiently safe to add to drinking water cisterns in mosquito-prone areas. This has helped to control malaria and reduce the spread of West Nile virus. It is also used to control fleas on domestic animals, and as a food additive in cattle feed to prevent fly-breeding in manure.

The opposite approach is to prevent the production of juvenile hormones by inhibiting biosynthetic enzymes. By inhibiting the production of juvenile hormones, moulting takes place too soon and produces non-functional adults. IGRs that inhibit the enzymes JH acid methyltransferase and cytochrome P450 CYP15

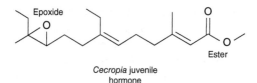

Cecropia juvenile
hormone

68. Example of a juvenile hormone.

are of particular interest. These enzymes are specific to insects, and so it should be feasible to design inhibitors that will have minimal side effects on mammals and other species.

IGRs that target the ecdysone receptor disrupt the transformation of larvae into adults at the pupal stage. Tebufenozide is an example of an ecdysone receptor agonist, and is effective in controlling caterpillars. The compound has high selectivity and low toxicity, earning the company that developed it (Rohm and Haas) a Presidential Green Chemistry Award.

Other IGRs inhibit the biosynthesis of chitin—an essential carbohydrate required for the exoskeleton—which means that insects are trapped in their old exoskeleton. These inhibitors are quicker acting and longer lasting than hormonal IGRs. One example is diflubenzuron, which was marketed in 1976 (Figure 69). Diflubenzuron is used primarily in forest management to control boll weevils and various types of moths. It is highly toxic to insects and relatively non-toxic to mammals.

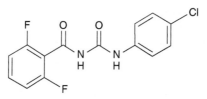

69. Diflubenzuron.

The search for new insecticides often involves taking a leaf out of nature's book. Several plants, fungi, and bacterial strains produce compounds that act as insecticides or repellents. For example, the bacterial strain *Bacillus thuringiensis* infects insects, and produces toxins that kill beetles, mosquitoes, and caterpillar larvae. Genetic engineering has been used to incorporate this bacterial toxin (Bt toxin) into plants.

Several plants emit volatile chemicals known as terpenes that act as insect repellents and could serve as the starting point for the design of new insecticides. One current research area involves synthesizing analogues of a natural product called germacrene D—a compound that repels aphids and other insect pests. A Japanese research team has also discovered that tomatoes exude a volatile chemical when they are under attack from caterpillars. This chemical acts as a chemical warning for neighbouring tomatoes, which then produce an insecticide to defend themselves against potential attack. Remarkably, this insecticide is produced from one of the alarm chemicals absorbed by the plant. It is possible that other plants have similar defence mechanisms and this could offer new approaches to the control of insects.

Fungicides

Fungicides kill fungal infections that are harmful to crops or farm animals. Some plants and organisms contain natural fungicides as a chemical defence against fungal disease. These fungicides include cinnamaldehyde, monocerin, cinnamon, citronella, jojoba, oregano, rosemary, and extracts from the tea tree and neem tree. The bacterium *Bacillus subtilis* and the fungus *Ulocladium oudemansii* can sometimes be used as fungicides, while kelp is fed to cattle to protect them from fungi in grass.

A number of synthetic fungicides produced in the laboratory have also proved effective. Examples of older fungicides include benomyl, vinclozolin, and metalaxyl. However, the more modern fungicides show better selectivity and potency. Examples of the latter include a class of compounds known as the 'quinone outside inhibitors' (QoI)—considered the most important development in fungicides in recent years. One example of this class of inhibitors is azoxystrobin (Figure 70), which was developed by Jealott's Hill International Research Centre from natural antifungal agents produced by a species of European small mushroom. The key feature for antifungal activity (the toxophore) is the portion encircled. Another example are the triazole fungicides—so called because they contain a triazole ring in their structure. Prothioconazole (Figure 70) is an example of this class of fungicide and has the added bonus that it stimulates plant growth.

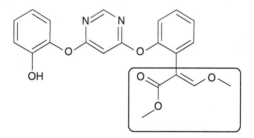

Azoxystrobin

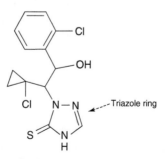

Prothioconazole

70. Examples of fungicides.

A co-formulation of fluoxastrobin (a strobilurin) and prothioconazole is commercially available under the trade name of Fandango and provides broader antifungal protection than using the two different antifungal agents on their own. Combining two fungicides that act on different targets lessens the chances of fungal strains gaining resistance. If resistance should arise to one of the fungicides, the fungal strain should still be susceptible to the other fungicide. For example, when metalaxyl was used to control potato blight in Ireland, resistance developed within one growing season. However, in the UK, resistance developed more slowly because metalaxyl was used in combination with other fungicides.

Resistance can arise due to mutations that alter key amino acids in the binding sites of target proteins. This often affects all the fungicides within a particular structural class—a property known as cross-resistance. For example, black sigatoka is a fungal disease of bananas that is resistant to all the QoI fungicides, and is caused by a mutation that replaces a glycine residue with alanine.

Herbicides

Herbicides control weeds that would otherwise compete with crops for water and soil nutrients. More is spent on herbicides than any other class of pesticide, with six billion dollars being spent in the USA alone. Common salt was used in historical times, while inorganic herbicides were used before World War II. However, these chemicals were not particularly selective and could damage crops.

Selective herbicides are necessary when treating crops, but non-selective herbicides are useful if the aim is to kill all plants on waste ground, industrial sites, and railways. Some plants produce natural herbicides that affect neighbouring plant life (a characteristic known as allelopathy). One example is the 'tree of heaven', which has been given less complimentary names such as the 'stink tree' or the 'tree from Hell'. This is because of its foul smell and invasive properties. Leaves of the black walnut tree contain a herbicide called juglone which is toxic to apple trees and a number of plants. When the leaves fall to the ground, the chemical is released and prevents other plants from competing for available space and nutrients.

A number of synthetic herbicides have been designed that mimic the action of plant hormones called auxins (Figure 71). These plant hormones are generated in response to external environmental conditions and coordinate plant growth. Natural auxins, such as 4-chloroindole-3-acetic acid, contain a carboxylic acid and an aromatic ring.

The synthetic agent 2,4-D (Figure 71) contains these same functional groups and mimics the action of auxins. It was synthesized by ICI in 1940 as part of research carried out on biological weapons, and was found to kill broad-leaved weeds without harming narrow-leaf cereal crops. It was first used commercially in 1946 and proved highly successful in eradicating weeds in cereal grass crops such as wheat, maize, and rice. It is one hundred times more potent than inorganic herbicides and was largely responsible for the post-war expansion in agricultural output. The compound is easy and cheap to synthesize, and it is still the most widely used herbicide in the world.

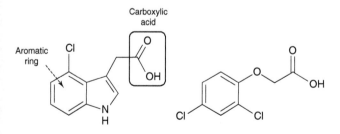

4-Chloroindole-3-acetic acid   2,4-D

71. Examples of auxins.

An ester derivative of 2,4-D was one of the active constituents present in Agent Orange—the herbicide used by US forces during the Vietnam war. In the 1950s, triazine herbicides such as atrazine (Figure 72) were developed (triazine refers to the presence of a six-membered aromatic ring containing three nitrogen atoms). These agents kill weeds by inhibiting a protein that is important in photosynthesis.

During the 1970s, a group of herbicides described as bleaching herbicides were introduced to the market. These act as enzyme inhibitors that prevent the biosynthesis of photosynthetic pigments. The first of these to reach the market was norflurazon in 1971 (Figure 72).

Another useful enzyme target for herbicides is acetolactate synthase—a key enzyme in the biosynthesis of amino acids such as valine, leucine, and isoleucine. Two groups of herbicides inhibit this enzyme (Figure 73). The sulfurons were developed in the 1980s and 1990s, and include chlorsulfuron. These agents have proved extremely potent. For example, only one ounce of chlorsulfuron (Glean) is required to treat one acre of land. The carbazones were developed more recently and include propoxycarbazone-sodium.

There are several other commercially available herbicides that act on different targets. One example is glyphosate, which is

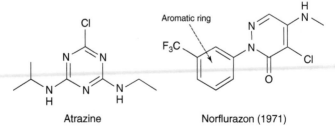

Atrazine   Norflurazon (1971)

72. Miscellaneous herbicides.

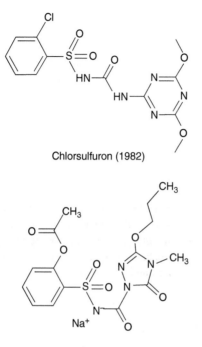

Chlorsulfuron (1982)

Propoxycarbazone-sodium (2001)

73. **Examples of acetolactate synthase inhibitors.**

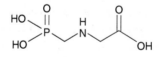

74. **Glyphosate.**

used in the garden weedkiller Roundup (Figure 74). This agent inhibits an enzyme involved in the biosynthesis of phenylalanine. Glyphosate is selective for weeds, because people and animals obtain phenylalanine from their diet and do not synthesize it. In other words, the target enzyme is not present in mammalian cells.

Chapter 7
The chemistry of the senses

Naturally occurring organic molecules play an important role in the way we perceive the world through the various senses of sight, taste, and smell. In addition, many synthetic organic molecules have been designed to have colour, taste, and scent, and are important in the food and cosmetic industries.

The chemistry of vision

A naturally occurring organic chemical called 11-cis-retinal is crucial to the mechanism by which the rod cells in the human eye detect visible light (Figure 75). Its structure includes a series of alternating single and double bonds, which is known as a conjugated system. A total of six double bonds are involved with one of the alkene groups defined as *cis* (see Chapter 2). This creates a kink in the chain. A reactive aldehyde group at the end of the side chain allows a chemical reaction to take place that results in a covalent bond being formed between retinal and a protein called opsin to produce a modified protein called rhodopsin. The conjugated system of retinal (defined as the chromophore) absorbs light and this causes the *cis*-alkene to become a *trans*-alkene, which straightens the chain. This, in turn, alters the shape of the protein and triggers a nerve signal that is transmitted to the brain, where it is interpreted as light.

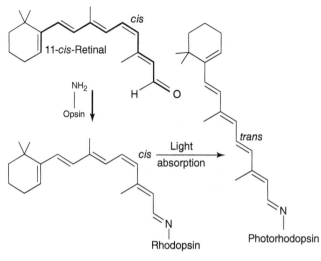

75. The role of retinal in the visual process.

The type of conjugated system present in a molecule determines the specific wavelength of light absorbed. In general, the more extended the conjugation, the higher the wavelength absorbed. For example, β-carotene (Figure 76) is the molecule responsible for the orange colour of carrots. It has a conjugated system involving eleven double bonds, and absorbs light in the blue region of the spectrum. It appears red because the reflected light lacks the blue component. Zeaxanthin is very similar in structure to β-carotene, and is responsible for the yellow colour of corn. Other naturally occurring pigments include lycopene and chlorophyll. Lycopene absorbs blue-green light and is responsible for the red colour of tomatoes, rose hips, and berries. Chlorophyll absorbs red light and is coloured green.

An understanding of how conjugated systems absorb light means that research chemists can design and synthesize coloured molecules. Of particular importance are a series of dyes that contain a diazo functional group (–N=N–) as part of the

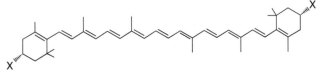

β-Carotene; X=H
Zeaxanthin; X=OH

76. Examples of molecules with conjugated systems.

conjugated system. Examples include the yellow dye tartrazine (Figure 77) and the red dye scarlet GN. Tartrazine is widely used as a food colouring (E102), and is also present in medicines, soaps, perfumes, toothpaste, shampoo, and moisturizers. Scarlet GN used to be approved as a red colouring in food (E125), but has now been replaced by alternative colourants. Several dyes have been banned as food colourings in recent years, but have been used illegally in some countries. For example, rhodamine B is banned as a food dye but was discovered in confectionary samples tested in India during 2013. Dyes are also important in colouring manufactured goods such as plastic, paper, cloths, and paints. For example, a diazo dye used for this purpose is Bismarck Brown Y.

Indigo (Figure 78) is an important natural dye that absorbs yellow light and is blue in colour. It can be extracted from plants, but it is much cheaper to produce it synthetically. Indigo has a long history and was used by the ancient Mayan civilization. It also has an important political and economic history. Indigo was a major Indian crop at the end of the 19th century, but disputes over the prices paid to Indian farmers led to Mahatma Gandhi proposing non-violent mass civil disobedience. The incident contributed to Gandhi becoming established as the leader of Indian nationalism. The unrest and disruption resulting from this episode also encouraged chemists to devise a synthetic process for indigo, with BASF setting up an industrial plant to produce the dye in 1897. It is claimed that this was the first industrial plant to produce a

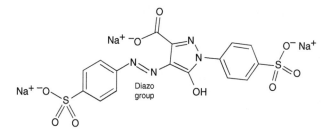

77. Tartrazine (E102).

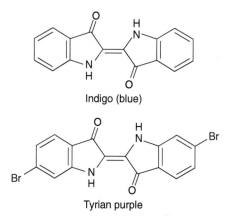

Indigo (blue)

Tyrian purple

78. Naturally occurring dyes.

synthetic compound. By 1910, all the indigo used in Europe was synthetic and imports from India had ceased.

Tyrian purple (Figure 78) is another naturally occurring dye that is very similar in structure to indigo and can be extracted from sea snails. It was used in Roman times to colour the robes of emperors and senators. Dyes such as Tyrian purple were extremely expensive and could only be afforded by the rich and the powerful. It was not until the advent of cheap synthetic dyes in the 19th century that colourful clothing became affordable to all.

Research into coloured molecules still takes place. One research team at Yale University has investigated fossils to try and identify the colour of dinosaurs. An organic compound called melanin has played a crucial role in this research. Melanin is a colour-determining pigment found in hair, skin, feathers, and fur. A study of current animals shows that the pigment is stored within cells in small containers called melanosomes. These 'cellular ink pots' have also been observed in fossils using a scanning electron microscope, and provide clues to what colour the fossil might have had. Sausage-shaped melanosomes commonly contain a molecule called eumelanin which is dark brown or black, whereas spherical melanosomes contain pheomelanin which is red. By studying the shapes of melanosomes in modern birds and comparing them with the bird's actual colour, it has been possible to propose colour patterns in feathered dinosaurs. Other fossil studies have detected eumelanin, which has survived for 160 million years.

A number of coloured molecules have useful medicinal properties. For example, a dye called prontosil was found to have antibacterial properties, and this led to the discovery of the sulphonamides in the 1930s. Another example is proflavine which has been used as a topical antibacterial agent. More recently, it has been discovered that microbial cells can be killed using a dye called methylene blue. Methylene blue is normally non-toxic, but becomes a photosensitizing agent when exposed to light. This causes reactive oxygen species to be produced which prove lethal to cells. The use of methylene blue and light to treat bacterial infections is an approach known as photodynamic therapy. It is highly selective in killing bacterial cells rather than human cells, since methylene blue is taken up more quickly by bacterial cells. Methylene blue contains a positive charge and is attracted to bacterial cells because they have a greater negative charge on their surface compared to mammalian cells. The treatment works best for surface infections that are accessible to light. Photodynamic therapy has also been used since the 1980s to treat some cancers.

Dyes are also being considered in the production of dye-sensitized solar cells (DSSCs), which work because electrons are produced when photons strike the dye molecules. DSSCs offer potential advantages over silicon-based solar cells. They can work under a variety of light conditions and do not need bright sun. They also work well with artificial light or with light coming from different directions at the same time. This means that DSSCs could potentially produce greater quantities of energy than conventional solar cells. Because the cells work well in low light, they could be used to power small electronic devices by harvesting indoor light. Recently, the MGM Grand Hotel in Las Vegas replaced its room curtains with remote controlled electric blinds powered by DSSCs.

Light-sensitive molecules are thought to influence the ability of birds to migrate and navigate accurately over thousands of miles. For example, bartailed godwits fly 10,000 km from Siberia to New Zealand each autumn, and can modify their flight even if they are blown off course. It is thought that birds can visually detect the Earth's magnetic field lines since their navigational ability is light-dependent. Photoreceptor proteins called cryptochromes have been discovered that might be responsible for an ability to measure longitude. However, the mechanism by which these molecules detect magnetic fields is still to be determined. Magnetite is also involved in direction-finding and may be responsible for determining latitude. Based on this research, a synthetic molecule has been designed that can respond to weak magnetic fields of similar strength to the Earth's magnetic field. It consists of a highly conjugated carotenoid, a porphyrin, and a fullerene (Figure 79).

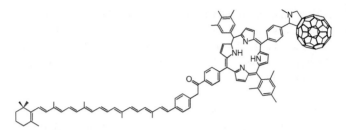

79. A synthetic molecule designed to detect magnetic fields.

The chemistry of scent

Scented molecules interact with olfactory receptors in the nose.
These receptor proteins were identified in 1991 by Richard Axel
and Linda Buck, earning them the 2004 Nobel Prize in Physiology
or Medicine. Scented molecules are important in the production
of a wide range of cosmetics and household goods, such as
perfumes, soaps, toothpaste, and detergents. Examples of organic
molecules used as scents include *cis*-jasmone (extracted from
jasmine flowers) and damascenone, which is responsible for
the smell of roses. Muscone is used to provide a musky aroma
in perfumes, while citral has a lemon fragrance. The two
enantiomers (mirror images) of a chiral molecule sometimes
have different scents. For example, the *R*-enantiomer of limonene
smells of oranges, while the *S*-enantiomer smells of lemons.
One of the enantiomers of carvone smells of spearmint while
the other smells of caraway.

Scented molecules play an important role in the natural world.
For example, insects exude pheromones that act as sex attractants
for potential mates, or as chemical alarm signals. There are a
large variety of pheromone structures, and they are extremely
potent. Some are effective at levels as low as 2×10^{-12} g. Since
pheromones are present in such small quantities, it would be a
huge undertaking to extract them from their natural sources. For
example, it requires 65,000 female cigarette beetles to extract

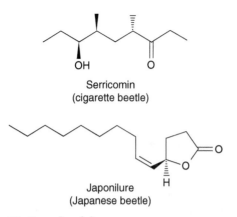

Serricomin
(cigarette beetle)

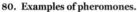

Japonilure
(Japanese beetle)

80. Examples of pheromones.

1.5 mg of the sex pheromone serricornin (Figure 80). Fortunately, most pheromones can be synthesized relatively easily in the laboratory. One commercial application is to use pheromones in traps to eliminate destructive insects. For example, Japonilure is marketed to trap beetles. Only 25 µg is required to catch thousands of the insects. Pheromones have also been used to trap boll weevils, flies, termites, and fruit moths.

Plants and mammals also have pheromones. For example, androstenone is a pig-steroid pheromone that causes a receptive sow to adopt the stance for sex. The pheromone is commercially available as a spray, which farmers apply across the nose of a sow. This allows artificial insemination to be carried out more easily. Pheromones can be fast-acting. Insect sex attractants and alarm pheromones are highly volatile and have an immediate reaction. Trail pheromones are used by bees, ants, wasps, and termites to indicate food sources. They are less volatile, but more prolonged. Some pheromones act more slowly and have longer term effects. Queen bee substance is produced by the queen bee, and prevents the development of ovaries in worker bees. Therefore, only the queen produces eggs. If the queen dies, the lack of pheromone

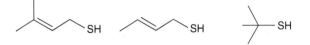

Thiols present in skunk spray   *tert*-Butyl thiol

81. Examples of foul-smelling thiols.

inspires worker bees to feed royal jelly to bee larvae in order to raise a new queen.

Pheromones are not the only scented molecules that are important in nature. For example, flowers use scent to attract bees, while molluscs can escape predator sea stars by detecting their scent. On the other hand, some predators detect their prey by scent. For example, the codling moth larva detects a chemical released by apple skins. Some natural scents act as repellents. For example, recent research has shown that golden orb spiders produce an ant repellent that could lead to the design of new insect repellents. The skunk emits foul smelling thiols when faced with danger (Figure 81). Such thiols have commercial uses. Since natural gas is odourless, *tert*-butyl thiol is added in order to detect gas leakages. The odour of this chemical is so powerful that it is present at only one part per 50,000,000,000 parts of natural gas.

Clearly, scented molecules have a commercial importance in perfumes, cosmetics, soaps, detergents, and air fresheners. Designing a perfume is as much an art as a science, since it involves combining different scented molecules to produce a unique scent that is quite distinct from the scents of the individual molecules involved. This has parallels in the natural world. The natural aroma of a rose is due mainly to 2-phenylethanol, geraniol, and citronellol. However, molecules such as damascone subtly influence the scent of each rose. Many scented molecules can be isolated from the natural world, but it is often easier and

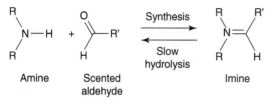

82. Slow release of a volatile scented aldehyde.

more ecologically friendly to synthesize them. For example, some of the scents present in flowers are present in such small quantities that many tons of flowers would have to be grown to extract them.

Synthetic chemistry can also produce novel scented molecules. For example, Chanel No 5 contains long-chain aliphatic aldehydes with aromas that are not found in the natural world. Chemistry can also influence how long a scent will last. For example, a highly volatile aldehyde would evaporate too quickly. Combining the aldehyde with an amine produces an imine that is less volatile and slowly hydrolyses to release the scented aldehyde over long periods (Figure 82).

Some chemicals used in perfumes and cosmetics can cause allergies for susceptible individuals. They include limonene, oak moss, and eugenol (present in cloves and spices). Limonene occurs naturally in citrus fruits, and it has been discovered that eating an orange at high altitude can sensitize skiers and mountaineers to limonene in cosmetics. North Americans commonly experience sensitization to urushiols in poison ivy. Unfortunately, urushiols are often present in lacquered toilet seats used in Asia, a fact that can have unfortunate results on some American tourists.

A number of current research projects involve volatile molecules. For example, understanding how mosquitoes 'sniff out' their prey could be useful in combating mosquito-borne diseases such as malaria and dengue viral fever. Carbon dioxide is the major attractant for mosquitoes, but various body odours have an influence. Some chemicals appear to mask chemical attractants and could serve as alternatives to DEET—the standard mosquito repellent. Several plants have been shown to attract mosquitoes, and the chemicals responsible (e.g. linalool oxide) have been used as mosquito traps. Alternatively, the plants could be grown round African villages to draw mosquitoes away from the village.

Research is being carried out into the design of sensors that detect volatile chemicals. Such 'electronic noses' could be used to detect chemical leaks in production plants, monitor the quality of food, or detect drugs or explosives. Sensors are also being developed to detect survivors of earthquakes or avalanches, or to locate dead bodies and secret graves. Sensors might also be used to determine how long a person has been dead, since the volatile organic compounds released from a dead body vary during different stages of decomposition. Bacteria emit volatile molecules, and detecting these could be a useful method of identifying which bacterial strain is causing an infection. If this proves reliable, it would speed up the time taken to identify the infection and provide the best antibacterial agents to treat it.

Finally, Luca Turin at the University of Ulm in Germany has proposed a novel theory regarding the mechanism by which scented molecules are detected by olfactory receptor proteins. Normally, a molecule activates a receptor because of its shape and binding interactions. Turin has suggested that bond vibrations might be more important. Turin's theory might help to explain why molecules with different structures have similar smells. For example, cyanide and benzaldehyde both smell of bitter almonds. The theory could also explain why similar looking molecules have significant differences in smell. On the other hand, critics have pointed out that there are around 400 different olfactory protein receptors in humans, and that a more complex biological process may be involved where different receptors are activated by a particular molecule. The brain then interprets the pattern of signals received from those interactions to detect a specific odour.

The chemistry of taste

The sensation of taste is caused by organic molecules interacting with the tongue's taste receptors. Different organic compounds can have different tastes, and may be used in the food industry to enhance flavour. A number of naturally occurring compounds are used in this way. For example, carvone is used in spearmint chewing gum and toothpaste to give a minty flavour. Other flavourings include menthol from spearmint oil, vanillin from vanilla, and benzaldehyde from almonds.

Synthetic flavourings are normally assumed to be non-natural compounds, but this is not necessarily true. Several natural flavourings such as vanillin are more conveniently synthesized in the laboratory.

One of the biggest markets for synthetic flavourings is in artificial sweeteners such as saccharin and aspartame (Figure 83)—a market worth £5 billion. Artificial sweeteners were introduced as low-calorie alternatives to sugar (sucrose) in order to counter the problems of obesity and diabetes. They are more potent than sugar, and can be tasted at lower concentrations, but this does not mean they have the same type of sweet taste experienced with sugar. Saccharin was the first synthetic sweetener and was used during World War I because of sugar shortages. It is 300 times more potent than sucrose. Cyclamate was discovered in 1937, and produces a better taste when it is combined with saccharin. Aspartame, sucralose, and neotame then followed. Aspartame is prepared from two naturally occurring amino acids (aspartic acid and phenylalanine), and is the most commonly used artificial sweetener today. It is 200 times more potent than sugar and can produce a similar taste to sucrose when combined with other sweeteners.

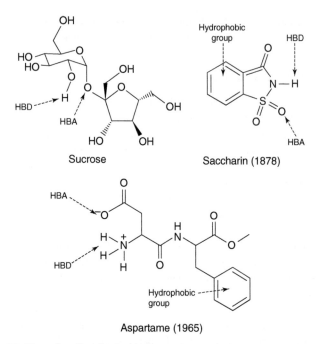

Sucrose

Saccharin (1878)

Aspartame (1965)

83. **Examples of synthetic sweeteners.**

Sucrose, saccharin, and aspartame have very different structures and so it is not obvious why they should all taste sweet. However, all three compounds contain a hydrogen atom that can act as a hydrogen bond donor (HBD), and an oxygen atom that can act as a hydrogen bond acceptor (HBA), separated from each other by about 3Å (0.3 nm). It is proposed that these groups form similar hydrogen bonds to sweet taste receptors.

The 'sweetness triangle' is an extension of this theory (Figure 84). The triangle defines three key binding regions within the receptor binding site. Two of these regions form hydrogen bonds, while the third region is hydrophobic and forms van der Waals interactions.

Organic Chemistry

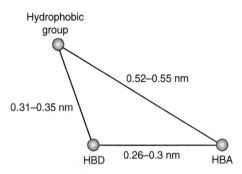

84. The sweetness triangle.

Molecules that can interact simultaneously with all three regions are likely to be sweet.

There have been various public debates over the safety of artificial sweeteners. Cyclamate was banned in the USA in 1969 because of concerns over possible carcinogenic effects, but it has been approved by the European Food Safety Agency (EFSA). Questions over the safety of saccharin persisted from 1971 until 2001, when the FDA finally declared that it was safe. Aspartame has also been branded unsafe by a number of consumer groups on the basis of dubious toxicological experiments. In fact, the EFSA gave aspartame a clean bill of health in 2013.

Due to the controversies over artificial sweeteners, several food and drink companies have started producing products with naturally occurring low-calorie sweeteners. For example, Coca Cola Life contains natural sweeteners called steviol glycosides that are extracted from the leaf of a South American shrub. This has allowed the level of sugar to be decreased to 37 per cent of what is present in regular Coke. The most potent of the steviol glycosides is called rebaudioside A. Another sweetener called erythritol is added alongside to provide a taste that is closer to that of sucrose.

The mogrosides are another group of naturally occurring sweeteners, which are extracted from the monk fruit found in South-East Asia. The sweetest of these is mogroside A or ergoside.

A number of plant proteins have also been found to be natural sweeteners. These include thaumatin, brazzein, pentadin, and miraculin. Miraculin is a protein extracted from the miracle fruit—a berry from a West African plant which has the remarkable ability to make sour foods taste sweet. Miraculin binds tightly to the sweet taste receptor for about an hour, but fails to activate it. If a sour, acidic food is taken during that period, the pH in the mouth drops and causes the bound miraculin to change shape. By doing so, it activates the sweet taste receptor, and drowns out the sour taste that would normally result. Miraculin is approved as a food additive in Japan, but not Europe or the USA.

The structure of the sweet taste receptor was identified in 2001 and consists of two membrane-bound proteins. The portion of the protein dimer that binds sugars is called the 'Venus fly trap' domain due to the way it changes shape when it binds sugars. Sweet taste receptors have also been discovered in the intestine, where they regulate the absorption of sugar into the bloodstream. They will respond to both natural and artificial sweeteners, and this may explain why low-calorie sweeteners fail to help people lose weight. An alternative approach to dieting is now being considered, whereby molecules are designed to inhibit the sweet taste receptors.

At the opposite extreme of the taste scale are the foul-tasting molecules. These, too, have a commercial use. For example, the aptly named Bitrex is added to toxic household products such as toilet cleaners to deter children from drinking them. Bad-tasting chemicals are also produced by plants to deter animals and insects. Nicotine plays this role for tobacco plants.

Chapter 8
Polymers, plastics, and textiles

Over the last fifty years, synthetic materials have largely replaced natural materials such as wood, leather, wool, and cotton. Plastics and polymers are perhaps the most visible sign of how organic chemistry has changed society. Celluloid—an early example of a polymer—was invented in 1856, and was used to produce billiard balls, piano keys, and early movie film. In 1891, the first synthetic fibre (rayon) was discovered by accident when Louis Chardonnet spilt nitrocellulose and observed the formation of silk-like strands. In 1917, synthetic rubber was synthesized in Germany in response to the British naval blockade. However, the real explosion in polymer science occurred in the second half of the 20th century. It is estimated that the production of global plastics was 288 million tons in 2012, and that plastic consumption could be one billion tons by the end of the century. The most prevalent plastics are poly(alkenes) such as polyvinyl chloride (PVC), polystyrene, expanded polystyrene, and polyethylene terephthalate (PET).

Polymerization involves linking molecular building blocks (termed monomers) into long molecular strands called polymers (Figure 85). By varying the nature of the monomer, a huge range of different polymers can be synthesized with widely differing properties. The idea of linking small molecular building blocks into polymers is not a new one. Nature has been at it for millions of years using amino acid building blocks to make proteins, and

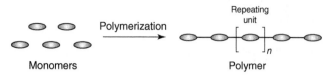

85. Polymerization.

nucleotide building blocks to make nucleic acids (Chapter 4). It has just taken scientists a little bit longer to mimic the process. Polymers have found practical uses as plastics, synthetic fibres, building materials, and adhesives. For example, nylon, polyesters, and polyacrylonitriles are commonly used for clothing. Other polymers used in clothing include Lycra, which has elasticity, and Dyneema which is the strongest fabric commercially available.

There are two general approaches to preparing polymers:

- Addition polymers (or chain-growth polymers)
- Condensation polymers (or step-growth polymers)

Addition polymers are formed from monomers such as alkenes, dienes, and epoxides. Each monomer is added one after the other to the end of the growing polymer chain and there are no atoms lost as a result of each addition. For example, when the monomers are alkenes, all the atoms of the alkene are incorporated into the polymer chain (Figure 86). However, the double bond is no longer present as one of the bonds was used in the linking process.

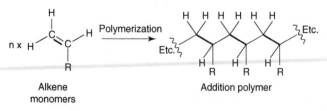

86. Example of addition polymerization. A bold bond has been included to highlight the monomers in the addition polymer.

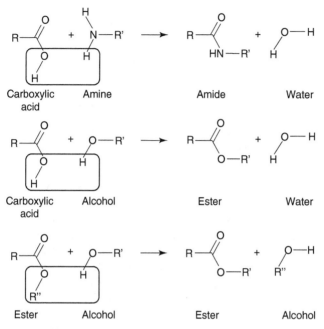

87. Examples of condensation reactions involved in polymerization.

Condensation polymers are formed when monomers are added by means of a condensation reaction. A condensation reaction is one which results in the loss of a small molecule such as water. For example, the reaction of an amine with a carboxylic acid to form an amide is an example of such a reaction. So too is the reaction between an alcohol and a carboxylic acid to form an ester. The reaction of an alcohol with an ester to form a different ester is also counted as a condensation reaction, although an alcohol is lost, rather than water (Figure 87).

Addition polymers

Polythene is also known as polyethylene or poly(ethene) (Figure 88). It was first produced by ICI in the 1930s as an electrical insulator

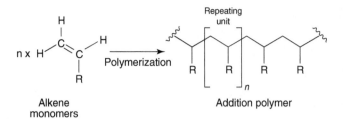

88. Addition polymers where R=H for polythene, R=CH$_3$ for polypropylene, and R=Cl for PVC.

for underwater telephone and telegraph cables. During World War II, its insulating properties were crucial in fitting radar sets to British planes. However, an efficient synthesis of the polymer was not perfected until 1953 when catalysts allowed the polymerization to be carried out under mild conditions. Several months later, a catalyst was used to produce polypropylene (also known as polypropene). Perfecting these syntheses earned the scientists involved the 1963 Nobel Prize in Chemistry.

Unlike polythene, polypropylene has substituents along the length of the polymer chain. The relative stereochemistry of these substituents plays an important role in the properties of the polymer. If the methyl groups point in opposite directions, then the polymer forms a regular helical structure that results in high fibre strength. If the methyl groups are randomly orientated, the polymer is rubbery and is of little commercial use. Polythene, however, was the polymer that sparked the rapid expansion of the polymer industry.

When addition polymers are prepared from alkene monomers, one of the bonds making up the double bond is used in the linking-up process. As a result, the resulting polymer is a fully saturated hydrocarbon chain containing no double bonds. When ethene is the monomer, there are no substituents in the final polymer (R=H). With other alkenes, there are regularly spaced substituents (R), and the nature of those substituents (R) influences the properties of the final product. For example, polymerizing the monomer vinyl chloride (R=Cl) gives poly(vinyl) chloride (PVC), which is used for plastic bottles, pipes, and clear food packaging. It is estimated that thirty-four million tons of PVC are produced each year, making it the third most produced plastic in the world.

Polymers have also been prepared from alkenes containing two or more substituents. For example, Teflon is prepared from a gas called tetrafluoroethene (Figure 89). It was first discovered by a scientist who came upon a full cylinder of tetrafluoroethene that provided no gas when the valve was opened. Intrigued, he cut the cylinder open and found that the contents had polymerized to produce a polymer that could not be melted and was inert to almost every chemical tested. Teflon is used as the coating for non-stick frying pans.

In theory, polymerization should produce very long polymer chains with no branches. In fact, this is not always the case. Sometimes a monomer links part-way along the polymer chain to create a branch. Branching can cause significant changes in the properties of the polymer. Unbranched, linear chains are able to pack together more closely than branched chains and produce a hard plastic. For example, unbranched polyethylene is a hard plastic that can be used in artificial hip joints. In contrast, polyethylene containing lots of branches is a flexible plastic used for rubbish bags and food bags.

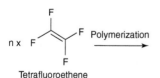 Polymerization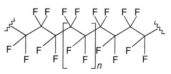

Tetrafluoroethene   Poly(tetrafluoroethylene)
                                            Teflon

89. Teflon.

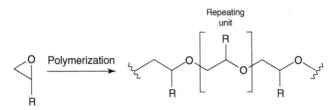

90. Addition polymers from epoxide monomers.

Addition polymers can also be prepared from epoxides—a three-membered ring containing an oxygen atom. The polymerization process opens up the epoxide ring of each monomer, and the oxygen atom is incorporated into the polymeric backbone to provide a polyether (Figure 90). Such polymers have found uses in skin-care products and pharmaceuticals, and as food additives.

Various rubbers are made from monomers containing a diene functional group (Figure 91). This is a group where two alkene groups are linked by a single bond. One of the double bonds is used for the linking process while the other shifts position. The original synthetic rubber (R=H) differed from natural rubber (R=CH₃) by lacking methyl substituents. Other synthetic rubbers have different substituents. For example, neoprene contains chlorine substituents and is used for wetsuits and coated fabrics.

The rubbers prepared in Figure 91 tend to be soft. In order to achieve a harder form of rubber that is more suitable for car tyres,

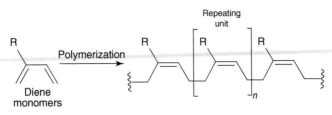

91. Polymers prepared from dienes.

it is necessary to carry out a process called *vulcanization*, where the rubber is heated with sulphur. This results in disulphide cross-links that hold the polymer chains together while still allowing stretching. The inventor of vulcanization (Charles Goodyear) discovered the process by accident when he spilt a mixture of rubber and sulphur on a hot stove. The more disulphide cross-links that are present, the harder the rubber. Thus, the rubber in elastic bands has a smaller number of cross-links compared to the rubber used in car tyres.

Addition polymers, called copolymers, can be synthesized from two or more different monomers. It is possible to control the polymerization, such that the monomers are introduced in a particular pattern. For example, it is possible to create linear polymers where the monomers are alternating or in blocks. It is also possible to create polymers where one chain containing a single type of monomer is grafted on to a second chain containing a different type of monomer (Figure 92).

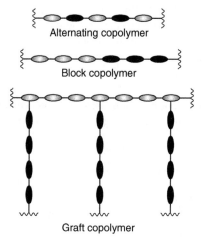

Alternating copolymer

Block copolymer

Graft copolymer

92. Copolymers.

Examples of copolymers include Saran, which is used as plastic film for wrapping food; styrene-acrylonitrile resin (SAN), which is used for dishwater-safe objects; acrylonitrile butadiene styrene, which is used in crash helmets; and butyl rubber, which is used in inner tubes and inflatable sporting goods. Copolymers are also being investigated in preparing nanocontainers that could be used in drug delivery.

Condensation polymers

Condensation polymerization requires two functional groups to be present on each monomer. In the natural world, protein biosynthesis involves condensation reactions, with each amino acid monomer providing an amine and a carboxylic acid. The synthesis of nylon can also involve amino acid monomers (Figure 93). Nylon was first introduced in 1939 and has been used for textiles, carpets, mountaineering ropes, and fishing lines. Different forms of nylon can be produced by varying the chain length of the monomer.

Some nylons are formed from condensation polymerizations involving two different monomers. For example, nylon 66 is a form of nylon prepared from one monomer containing two amine groups, and a second monomer containing two acid chloride groups (Figure 94).

Kevlar is another polymer formed from two different monomers (Figure 95). It is five times stronger than steel, and is used in space

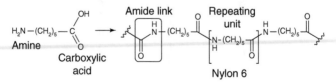

93. Synthesis of nylon 6.

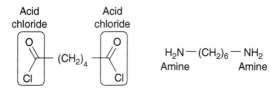

94. Monomers used to produce nylon 66.

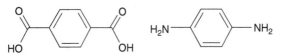

95. Monomers used for Kevlar.

suits, army helmets, bulletproof vests, and sporting equipment. Its stability to very high temperatures also means that it is used in protective clothing worn by firemen.

The remarkable strength of Kevlar is related to a number of factors. To begin with, each polymer chain is relatively rigid due to the planar aromatic rings and the limited bond rotation of the amide groups connecting each ring. Second, there is an extensive hydrogen-bonding network between the chains where each oxygen acts as an HBA and each NH proton acts as an HBD (Figure 96). This holds the polymer chains together into a rigid sheet and prevents the chains from slipping past one another. Finally, when Kevlar is spun into fibres the polymer chains are orientated along the fibre axis. This allows flat sheets of Kevlar to be stacked in a highly crystalline structure.

Condensation polymers linked by ester bonds are known as polyesters and are used for a variety of purposes including clothing. For example, Dacron is a polyester prepared from a diester and a second monomer containing two alcohol groups (Figure 97). Mylar is a similar polymer that is tear resistant and is used for sails and magnetic recording tape.

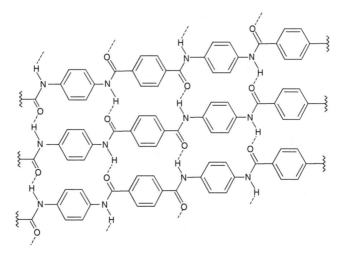

96. Intermolecular interactions between the polymer chains of Kevlar (hydrogen bonds are indicated by the dashed lines).

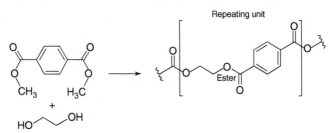

97. Condensation polymerization to form Dacron.

Polycarbonates are condensation polymers containing carbonate links. They provide a clear, transparent plastic that is lightweight, shatter resistant, and stable to heat. One example is Lexan (Figure 98), which is used for bulletproof windows, and compact discs. The polymer is made from diphenyl carbonate and bisphenol A.

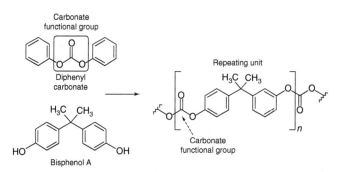

98. Condensation polymerization to form Lexan.

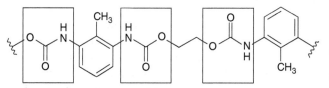

99. Structure of a polyurethane. The boxed regions define the urethane linkers.

Polyurethanes contain urethane functional groups as linkers (Figure 99). Polyurethane foams are used in furniture, bedding, and insulation. Polyurethanes are also used in fabrics such as Lycra.

Epoxy cements and superglues

When superglue and epoxy cements are used to stick surfaces together, a chemical reaction takes place to produce a cross-linked polymer (Figure 100). A polymer that contains epoxide rings at the end of each polymer chain is applied to both surfaces. A hardener is then applied which consists of a monomer containing two amine groups. The amine groups react with the epoxide rings to create the cross-links that stick the surfaces together. Polymers similar to superglue are being used to close wounds in surgery.

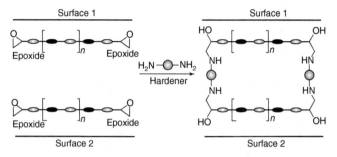

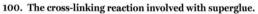

100. The cross-linking reaction involved with superglue.

Health issues

In recent years, safety concerns have been raised over several polymers, or the monomers used to make them. For example, long-chain polyfluorinated polymers such as Teflon are under particular scrutiny. These polymers are very stable and repel liquids, so they are useful in non-stick frying pans, raincoats, and surfactants in food packaging. They are also applied to skis in order to increase slip. However, they are now considered dangerous for health and harmful to the environment because of their persistence and bioaccumulation. The polymer industry argues that using shorter chains of these polymers would address many of the concerns raised, but critics disagree.

Health concerns have also been raised over polycarbonates and epoxy resins because of the monomer bisphenol A (BPA) used in their preparation. Polycarbonates are used in plastic bottles, while epoxy resins are used to coat the inside of food tins, drink cans, bottle tops, and water pipes. The coating protects metal from corrosive foods such as tomatoes, and also prevents food or drink taking up a metallic taste. However, trace levels of unreacted BPA monomer have been detected in these plastics and consumer groups are worried that it might leach out into food. As yet, there is no evidence that BPA is toxic to humans, but it is known to mimic the hormone oestrogen in animal studies.

Several countries have now banned the use of BPA-derived polymers in baby bottles since the addition of boiling water might encourage the monomer to leach out. The French government also decided to outlaw its use in food packaging and medical devices. There have also been some bans on its use in toy packaging. However, other countries have questioned the relevance of the toxicity tests carried out on BPA. For example, unrealistically high levels of BPA have been used in several toxicity tests.

Polyesters and polypropylenes can be used in place of polycarbonates for baby bottles, and a polyester called Tritan has been particularly useful in this respect. Finding alternatives to epoxy resin coatings for food containers is not so straightforward. One possible alternative is a compound based on lignin, which is a by-product of the paper industry. Two breakdown products of lignin have been combined to produce bisguaiacol-F (BGF) (Figure 101), which is structurally related to BPA but lacks its endocrine properties. The polymers derived from BGF have similar properties to those derived from BPA, and it is anticipated that BGF-based plastics will be available in under five years' time.

Environmental, ecological, and economic issues

Plastics have had beneficial ecological effects by reducing the use of natural materials for consumer goods. In the 19th century, celluloid replaced ivory, thereby reducing the slaughter of

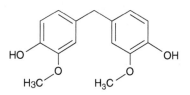

101. Bisguaiacol-F (BGF).

elephant herds. Today, synthetic fibres are used for clothes rather than cotton, and so fields that were once used to grow cotton are now used for food crops.

Another benefit of polymers is their stability and durability, allowing the production of sturdy, unbreakable products. However, this can have detrimental environmental effects when those products are carelessly discarded. The world's shorelines bear sad witness to this, and a report issued in 2015 estimated that eight million tons of plastic waste ends up in the oceans each year. On land, the pressure on landfill sites has become so intense that plastic recycling is fast becoming a necessity. There are now EU proposals to phase out the landfilling of recyclable waste by 2020, to discourage incineration, and to remove dangerous plastics from the market. EU member states are also required to clamp down on the use of thin plastic carrier bags. In 2010, an estimated eight billion of these bags ended up as litter.

Recycling/depolymerization

The raw materials for plastics come mainly from oil, which is a finite resource. Therefore, it makes sense to recycle or depolymerize plastics to recover that resource. Virtually all plastics can be recycled, but it is not necessarily economically feasible to do so. Traditional recycling of polyesters, polycarbonates, and polystyrene tends to produce inferior plastics that are suitable only for low-quality goods.

Depolymerization is possible using ruthenium, rhodium, or platinum catalysts, and the resulting monomers can be purified and reused for a variety of applications. However, depolymerization is not yet economically viable.

Biodegradable plastics

Efforts have been made to develop biodegradable polymers which are broken down by microorganisms. For example, polylactides are formed from lactic acid (Figure 102) and are used for food packaging, fabrics, and medical applications. Polyhydroxyalkanoates are biodegradable plastics that can be used instead of polypropylene. In the future, plant starches might be used to produce biodegradable polymers that would replace polycarboxylates.

Oxo-biodegradable plastics are conventional plastics that have been produced containing small levels of metal salts. These serve to catalyse biodegradation of the plastic when it is exposed to oxygen, and leads to microplastics which vary in size from 1nm to 5mm. It was hoped that these microplastics would be completely degraded by microorganisms, but this has not yet been proven. Moreover, the risks of microplastics on human health and the environment have not been fully investigated. There are concerns that microplastics could affect the food chain if they are ingested by marine fauna, and the EU is now considering the introduction of regulations on the use of microplastics in cosmetics, detergents, and other products.

Polylactides

Polyhydroxalkanoates

102. Biodegradable polyesters.

Bioplastics

Bioplastics are produced from monomers obtained from plant material rather than oil. In 2009, Coca Cola developed recyclable PET plastic bottles made partially from plant material. This involved the use of monoethylene glycol derived from sugar cane, but there are also plans to use fruit peel, bark, and stalks as sources for the chemical. Pepsi have made PET bottles using sources such as switchgrass, pine bark, and corn husks. Using plant material that would normally be discarded or burnt is preferable to using food crops, since the latter option would reduce food production and drive up food prices. In 2007, there were riots in Mexico because of high food prices resulting from corn being used for the production of ethanol as a biofuel. A variety of monomers have now been obtained from biological sources. Furan-2,5-dicarboxylic acid is one such example and has been polymerized to make poly (furan-2,5-dicarboxylic acid) (PEF)—a furan version of PET.

Recent research in plastics and polymers

Novel polymers are continually being developed for novel applications. For example, self-healing polymers are being developed that repair any physical damage to the material containing them. One approach is to incorporate miniature capsules, some of which contain a monomer and others a polymerization initiator. When damage occurs, the capsules at the damage site break to release their contents. The monomer and polymerization initiator mix, and a polymerization reaction takes place that repairs the damage. A different approach is to design polymers that depolymerize under certain conditions. In this case, a damaged area is exposed to light or heat to initiate depolymerization. The resulting monomers are more fluid than the polymer and fill any gap or tear that is present. When the light or heat is removed, the monomer repolymerizes to repair the damage.

Polymers are being used in the design of smart clothing. For example, polymers are involved in the design of thin-film thermoelectric devices that could be incorporated into such clothing. These would use body heat as an energy source to provide sufficient power for electronic devices such as mobile phones. Chemical sensors involving polymers are being developed to detect explosive vapours at concentrations of parts per quadrillion (1 in 10^{15}). The polymer turns red in the presence of airborne or waterborne TNT, and the system could be used in smart clothing to alert the wearer to the presence of explosives in former warzones contaminated with land mines or explosive weapons.

A UK company called Econic Technologies is developing a polymerization process that captures carbon dioxide emissions by reacting CO_2 with epoxides (Figure 103). Therefore, each carbonate group in the polymer would incorporate one captured molecule of CO_2. A polyurethane capable of absorbing methane—another greenhouse gas—has also been developed. There are huge reservoirs of methane under the ocean floor and concerns have been raised that global warming might cause some of this to be released. Another area of research involves porous organic polymers that bind toxic chemicals. These could prove useful in gas masks.

Biodegradable polymers have applications in medicine in the form of dissolvable stitches, plates, screws, pins, and meshes. Biodegradable polymers are also being investigated that could

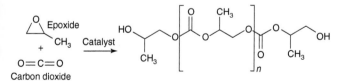

103. Carbon dioxide capture by polymerization.

potentially replace the titanium-based implants that are currently used to repair fractures in load-bearing bones.

Thermoresponsive hydrogel roof coverings are being developed that could absorb rain water then 'sweat' it out to help cool a building, thus cutting down on air conditioning and CO_2 emissions. The polymer being investigated is poly(N-isopropylacrylamide) (PNIPAM), which can store 90 per cent of its weight in water.

Finally, new polymers are being developed that would prevent chewing gum sticking to clothes, pavements, or floors. The polymer absorbs water, which promotes degradation, but which also gives the gum a longer lasting flavour. The polymer could also be used in lipsticks and lip balms, and is being considered as a method of treating bad breath.

Chapter 9
Nanochemistry

Nanochemistry involves the synthesis of molecular nanostructures measuring 1–100 nm. These could serve as molecular components for nanorobots and other molecular devices that could be used in medicine, analysis, synthesis, electronics, data storage, or material science. One current research goal is to design a molecular computer. Current computers use silicon integrated circuits, but there is a limit to how small these components can be made. Designing electronic devices and computers that operate at the molecular level will allow a dramatic reduction in scale and a corresponding increase in computer power. In order to make these dreams a reality, it is necessary to design nanostructures that are the molecular equivalents of wires, switches, data storage systems, and motors.

Carbon allotropes

Allotropes are ordered structures consisting entirely of one type of atom. Diamond is a carbon allotrope where each carbon atom is covalently linked to four other carbon atoms to form a very strong lattice (Figure 104a). Because of its strength, diamond has industrial applications such as diamond-tipped mining drills. It is one of the hardest and most chemically inert materials known, and it also has useful optical properties.

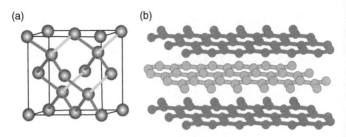

104. Structures of (a) diamond and (b) graphite

Graphite is another carbon allotrope where the atoms are arranged in layers of planar, aromatic rings (Figure 104b). Each layer involves strong covalent bonds, but only weak intermolecular interactions exist *between* each layer. This allows the layers to 'slide', which makes graphite ideal as the 'lead' for pencils. For the same reason, graphite is used as a dry lubricant in machinery and engines. Surprisingly, graphite has been detected in hip transplants, where it seems to lubricate the metal–metal contact of the hip joint. It is not yet known how this graphite is generated, but it is possible that the hip implant itself 'grinds up' proteins to form carbon, which is then converted to graphite.

Graphite also conducts electricity because of the relatively mobile π electrons present in aromatic rings. A recent innovation that makes use of graphite's conductivity involves the attachment of three different enzymes to graphite beads. One of the enzymes catalyses the splitting of hydrogen gas to form two protons and two electrons. Because of graphite's electrical conductivity, the electrons shuttle through the bead to a second enzyme which catalyses the reduction of a substrate. The product from that reaction undergoes a further reaction catalysed by the third enzyme. This system can be viewed as a miniature chemical factory and gained an Emerging Technology Award from the Royal Society of Chemistry in 2013.

A single layer of graphite is called graphene, and was first produced in 2004 at the University of Manchester, earning its inventors the 2010 Nobel Prize in Physics. As well as conducting electricity, graphene is the thinnest, strongest material known to science with a tensile strength 300 times greater than steel. It is also stable to heat, and relatively inert to chemicals.

Because of these properties, graphene has many potential applications as a component in chemical sensors, medical devices, solar cells, hydrogen fuel cells, batteries, flexible displays, and electrical devices. One potential use for graphene is as a desalination filter. The idea would be to punch pores into the graphene that would allow water to squeeze through, but not salts. Another possible application for graphene is as an alternative to Kevlar in body armour. In the field of sensors, it is thought that a graphene-based device might be capable of detecting bacterial infections or contamination.

At present, much of the work carried out on graphene has been in the research lab, and the next stage is to apply that research to create new materials in the factory—a process that may take 20–40 years. One practical problem is devising an economical method of producing graphene on a large scale. This is essential if it is to be used commercially.

Fullerenes are a third type of carbon allotrope, involving spheres or cage-like structures. The carbon atoms are arranged in hexagonal and pentagonal rings, the latter introducing the curvature required to form a sphere. The best-known example of a fullerene is buckminsterfullerene C_{60} (or fullerene-60), which has a pattern and shape akin to a soccer ball (Figure 105)—the number 60 refers to the number of carbon atoms present in the structure.

Fullerene-60 was discovered from experiments that were designed to mimic what kind of chemical reactions might be taking place

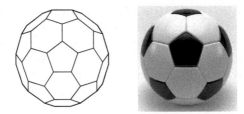

105. Arrangement of carbon atoms in fullerenes.

in outer space. In 2010, an infrared telescope established that fullerene-60 was indeed present in interstellar gas clouds. However, C60 structures are also formed much closer to home—near candle flames! It is thought that the process starts with the formation of small carbon cages which progressively increase in size by swallowing up atoms of vapourized carbon. However, it is not yet understood how the small cage fullerenes are formed in the first place. Fullerenes of different cage sizes include C28, C32, C50, and C70. Unlike C60, these are not perfectly spherical. For example, fullerene-70 is shaped like a sausage. The discovery of buckyballs earned Henry Kroto the Nobel Prize.

To date, fullerenes have not been put to any commercial use. However, there are plenty of suggestions for future applications. One suggestion is to use them as drug-delivery vehicles to introduce drugs or genes into cells. Other potential applications include lubricants, electrical conductors, solar cells, and even safety goggles. A modified fullerene has been used in the synthesis of a molecule that can respond to weak magnetic fields (Figure 79), and fullerenes have also been used as wheels in a nanocar (Figure 117).

Nanotubes

Carbon nanotubes are molecular cylinders made up of carbon atoms. The walls of the nanotube are made up of hexagonal rings

(Figure 106), and are essentially a rolled up layer of graphene. Each end of the nanotube is fullerene in nature and contains pentagonal rings that introduce the curvature that seals off the tube. Their diameter is about 1 nm (about the same diameter as a strand of DNA), and their length can be up to 132,000,000 nm.

The properties of nanotubes vary depending on their dimensions and atomic arrangements. Different lengths and diameters result in different electronic properties, which make nanotubes useful in nanoelectronic circuitry as insulators, semiconductors, or conductors.

The relative orientation of the rings making up the walls of a nanotube has a profound effect on its electronic properties. Hence, nanotube A in Figure 106 is a semiconductor, whereas nanotube B is a full conductor. The price of nanotubes is dropping dramatically and it is anticipated that they will have a widespread use in electronics by 2020. The field of nanoelectronics could well result in future molecular computers.

Nanotubes have been shown to be stronger than steel, but at a sixth of the weight, making them extremely useful in materials science. The strength to weight ratio of nanotubes gives them potential applications in aircraft parts, car parts, and sports equipment, especially if the nanotubes are 'bundled up' to form fibres with high tensile strength. Their large surface area is also

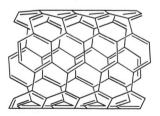 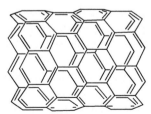

106. Structural variation in nanotubes.

useful. For example, they could be linked to enzymes and used for synthesis or in hydrogen fuel cells.

Nanotubes can be single-walled or multi-walled. Multi-walled nanotubes are multiple rolled layers of graphene and have improved chemical resistance. This is important when linking molecules to the surface of nanotubes, since the linking process can punch holes in the nanotube wall and affect its mechanical and electrical properties. With a double-layered nanotube, only the outer layer will be affected. Nanotubes have been modified to bind organic molecules capable of detecting other molecules, which make them useful components in sensors or bioelectronic 'noses' for monitoring food quality, or detecting explosives and chemical leaks. Alternatively, nanotubes linked to light-sensitive molecules could be useful in the design of solar cells and energy storage.

Nanotubes might also be useful as capsules for other molecules. Nature has already achieved this. The tobacco mosaic virus consists of a nanotube made up of identical viral proteins. The proteins self-assemble to form the nanotube and encapsulate viral RNA. Some researchers are looking into designing self-assembling nanotubes that will hold buckyballs, since it is believed that these will have good electronic properties.

Rotaxanes

A rotaxane is a nanostructure where two interlocking molecules form the equivalent of an axle and a wheel (Figure 107). The molecule representing the wheel is a large cyclic structure (macrocycle), while the molecule acting as the axle is dumbell-shaped. The two bulky groups at either end of the axle prevent the macrocycle 'slipping off' the axle. The macrocycle can either rotate around the axle or move along its length from one end to the other. However, the latter movement is not a smooth process since the axle contains one or more 'docking' sites that temporarily

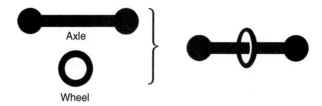

107. General structure of a rotaxane involving a dumbell-shaped molecule 'threaded' through a macrocycle.

hold the macrocycle in position. This is known as a molecular shuttle. The interactions are strong enough to ensure that the macrocycle spends most of its time interacting with the docking sites available, but weak enough to allow the macrocycle to shuttle between the available sites.

One example of a molecular shuttle is shown in Figure 108. The axle has two docking stations involving aromatic rings, and bulky silicon groups at each end to prevent the 'wheel' slipping off the axle. The aromatic rings at each docking site can interact with the aromatic rings of the macrocycle—a form of interaction known as a π–π interaction. The interaction is relatively weak and so it is possible for the wheel to shuttle between both docking sites. However, in this example there is a preference for the wheel to bind to the right-hand docking station. That is because the docking stations are not identical. One of the docking stations has oxygen atoms attached to the aromatic rings, while the other has nitrogen atoms. The latter site interacts more strongly with the 'wheel', and so the wheel spends 84 per cent of its time bound at that site, and the remaining 16 per cent bound at the other docking site.

This preference can be altered. Under acidic conditions, the nitrogen atoms attached to the right-hand docking site become protonated and gain a positive charge. Since the wheel already contains positively charged nitrogen atoms, it is repelled

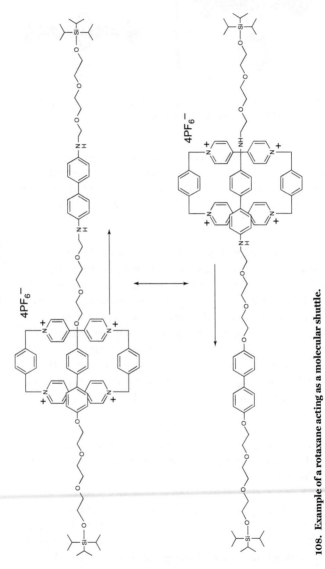

108. Example of a rotaxane acting as a molecular shuttle.

from the right-hand docking site and binds exclusively to the left-hand docking site. Therefore, the rotaxane acts like a molecular switch. It is not a perfect switch, since there should be exclusive docking to different docking sites under different conditions. Nevertheless, the example illustrates the potential of rotaxanes as molecular switches.

Molecular switches can have a number of applications. For example, a research team at Edinburgh University designed a rotaxane that acted as a 'switchable' catalyst for organic synthesis (Figure 109). The nitrogen atom at the centre of the axle is responsible for catalytic activity. Under basic conditions, the wheel binds to either of the two docking sites leaving the nitrogen atom free to act as a catalyst. Under acid conditions, the nitrogen becomes protonated and gains a positive charge, making it a stronger docking site for the wheel. The wheel now moves to the centre of the axle and conceals the catalytic site. In this example, the wheel contains oxygen atoms, which form strong hydrogen bonds with the protonated amine (Figure 110).

More recently, a rotaxane has been designed with two docking sites capable of catalysing two different kinds of reaction. The

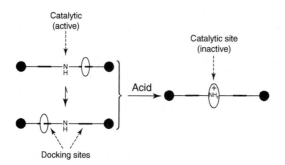

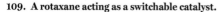

109. A rotaxane acting as a switchable catalyst.

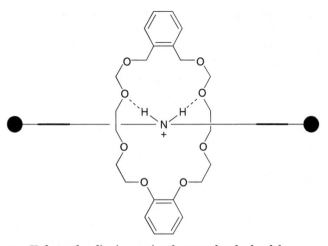

110. Hydrogen bonding interactions between the wheel and the protonated amine on the axle.

ring binds to one of the catalytic docking sites under acid conditions, and binds to the other catalytic site under basic conditions. Therefore, the same rotaxane can be used to catalyse two different kinds of reaction depending on the reaction conditions used.

Another rotaxane was designed to synthesize a tripeptide (Figure 111). The axle had three amino acids attached, and the wheel was threaded on to one end of the axle. As the wheel moved along the axle, it picked up the amino acids one by one in the order presented. There was no blocking group at the other end of the rotaxane and so the wheel with the attached tripeptide dropped off when it reached the end. The tripeptide could then be cleaved from the wheel. This research demonstrates that it is possible to design molecular synthetic machines that automatically produce new molecules, but there is a long way to go before this approach can compete with conventional synthesis.

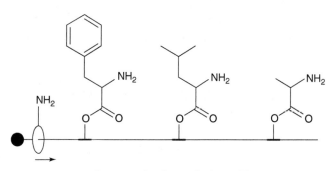

111. A rotaxane acting as a molecular synthetic machine.

112. Rotaxanes containing linear alkyne groups.

Rotaxanes with an axle made up of alkyne functional groups have been synthesized at Oxford University (Figure 112). Since alkynes are linear, the axle is linear and contains only carbon atoms. Such rotaxanes have been proposed as potential molecular wires for nanoelectronics. The wheel would act as an insulator as it shuttles back and forth.

Another approach towards molecular wires is to prepare polyrotaxanes that contain several wheels on the central axle (Figure 113). When several wheels are present on the one axle, they interact with each other, and this serves to stiffen and straighten the rotaxane. The efficiency with which electrons travel along molecular wires is faster with rigid rotaxanes.

If rotaxanes are to prove useful as switches or wires, they will have to be connected and integrated into rigid structures. One approach is to incorporate rotaxanes into the structure of metal-organic frameworks, such that the moving parts of each rotaxane is located within a pore. If this proves successful, then

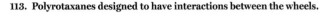

113. Polyrotaxanes designed to have interactions between the wheels.

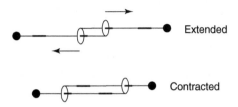

Extended

Contracted

114. Linked rotaxanes that mimic muscle action.

it opens up the possibility of creating solid-state molecular switches or machines.

Rotaxanes have also been used to create molecular 'muscles' that contract or expand under different stimuli. This involves two interlinked rotaxanes, where the end of each axle is covalently linked to the wheel on the other axle (Figure 114). This has been dubbed a daisy chain rotaxane. The lengths of the extended and contracted forms are 4.8 nm to 3.6 nm respectively. By polymerizing these daisy chain rotaxanes into molecular 'fibres', the resulting contractions and expansions are magnified (Figure 115). A French research team has linked together 3,000 rotaxanes that contract from 15.8 micrometres to 9.4 micrometres. The polymerization of the rotaxanes was achieved by using blocking groups that bind to a metal ion. Metal ions then act as a molecular 'glue' to hold the daisy chain rotaxanes together. The next challenge will be to bundle these fibres together.

Metal ion   Metal ion   Metal ion

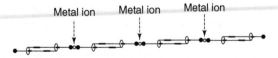

115. Polymerized daisy-chain rotaxanes into a molecular muscle fibre.

Nanoparticles

Nanoparticles are approximately 1–100 nm in size. Their properties are distinct from larger scale materials because of their size and relatively large surface area, and they have a wide range of actual and potential applications in medicine, manufacturing, materials, energy, and electronics. For example, it is possible to synthesize spherical nanoparticles that encapsulate a drug or DNA, then administer them to deliver their load to a patient's cells. For example, a lipid nanoparticle carrying the anti-cancer drug paclitaxel (Taxol) is currently undergoing clinical trials. Nanoparticles are also being designed that combine diagnostics with therapeutics (theranostics). For example, a nanoparticle has been designed to identify tumour cells. On binding, it breaks open to release an anti-cancer drug that treats the cancer, along with a dye that reveals where the tumour is.

Nanodelivery systems can also be used to protect neutraceuticals (e.g. vitamins) from the destructive effects of stomach acids. Nanocapsules have been constructed from proteins and sugars that are naturally found in food. The nanocapsules are stable to stomach acids, but are broken down by enzymes in the intestines to release the neutraceutical. Nanocapsules loaded with vitamin D could be added to soft drinks to prevent rickets.

Nanoparticles have uses in medicine, other than drug delivery. For example, nanoparticles have been developed that could potentially stop internal bleeding resulting from road accidents or terrorist bombs. The nanoparticles are designed to stick to activated platelets and speed up clot formation, thus reducing the chances of a patient bleeding to death. So far, the technique has only been tested on animals.

It has been discovered that carbon nanoparticles prevent the development of mosquito larvae, and so they could be useful in controlling malaria. The nanoparticles have a long lifetime, which would be an advantage in terms of their insecticidal activity, but could be a potential disadvantage if they have unforeseen environmental or ecological effects.

Nanotechnology and DNA

Nanostructures constructed from DNA have many potential applications. DNA is nature's data storage molecule and carries the codes required for an organism's proteins. In addition, its structure allows that information to be copied from one generation to another. The nucleic acid bases (ATGC) are the genetic alphabet, and a molecular recognition process takes place such that the base pairs are always A–T or G–C. This is crucial to the double helical structure of DNA, as well as the 3D shapes of RNA molecules.

Scientists have now taken advantage of base pairing to synthesize single-strand DNA molecules that self-assemble into predictable shapes determined by the sequence of bases present. For example, if a DNA strand contains complementary base sequences at different parts of the strand, then the molecule can coil up to allow base pairing (Figure 116). Using this approach, scientists have created 2D pictures using DNA, as well as 3D shapes. This is a process known as DNA origami.

This approach has been used to construct DNA nanorobots that can carry out robotic tasks such as sensing, computation, and cell-targeting. One research group has created a barrel-shaped DNA robot that is 35 nm in diameter and 45 nm long. The structure includes a hinge that allows the barrel to open up like a clam. Short DNA strands are present to keep the barrel closed until the nanorobot encounters an antigen that interacts with the DNA strands. This unlocks the barrel, which can then open to release its contents. So far, the nanorobot has only been tested on cell cultures, but it has the potential to carry drugs or

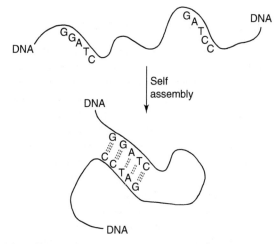

116. DNA origami.

antibodies to specific sites in the body. A similar idea for drug delivery involves DNA cubes that are designed to 'unzip' when they interact with RNA molecules that are unique to prostate cancer cells.

A DNA 'walker' has been designed that responds to light and can follow a trail along a surface. The trail consists of a series of 'poles' represented by strands of DNA, with each pole having a long section and a short section. The walker has two DNA legs—one short and one long. The walker binds to the first pole with its long leg binding to the long section of the pole and its short leg binding to the short section of the pole. When light is shone on the surface, the short section of the pole is split from the long section and floats away. The short leg of the walker is now free to seek out the short section of the next pole and bind to it. When it succeeds, it pulls the long leg along with it. In theory, this system could be used to design a nanolaboratory, where the walker picks up building blocks from different poles and combines them to form a product.

Examples of nanodevices and nanomachines

Nanodevices are being designed that mimic instruments or machines at a molecular level. For example, a nanodevice the size of a memory stick has been designed that can sequence DNA. The device makes use of two proteins. One of the proteins is a genetically modified form of a natural protein called α-hemolysin. This protein contains a pore and is embedded in a membrane-like surface, such that a nanopore is created through the membrane. The second protein can bind DNA and is linked to the outer surface of the pore protein. When it binds DNA it feeds it through the nanopore. As the DNA is threaded through the pore, the flow of ions through the pore varies depending on which base is in transit. The variation in ion flow can be measured and allows the DNA to be sequenced. Currently, the instrument can sequence up to 48,000 bases. A similar approach could potentially sequence proteins.

An all-carbon photovoltaic cell has been produced that involves carbon nanotubes, fullerenes, and graphene. The carbon nanotubes act as the light absorber and electron donor, while fullerene-60 buckyballs act as the electron acceptor. These are sandwiched between an anode of reduced graphene oxide and a cathode of more carbon nanotubes. The efficiency of the cell is too low to be commercially useful, but the technology could be incorporated into current solar cells to make them cheaper and more efficient.

A number of research teams have been involved in what might seem rather unusual projects, such as the design of molecular motorboats, cars, and trains. These may seem no more than curiosities, but the knowledge gained from such projects could eventually lead to commercially useful nanomachines. An example of a nanocar was synthesized in 2005 (Figure 117). The wheels are fullerenes, and a rigid molecule constructed from straight

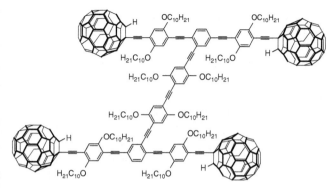

117. A nanocar.

chain aromatic rings and alkyne groups serves as the chassis. In truth, this contraption is better described as a nanocart, since there is no molecular motor present to propel it. However, research teams are working on that! The contraption can roll across a surface because the bonds linking the buckyball wheels to the chassis are rotatable.

Nanotechnology: safety and toxicology

Nanotechnology is already used in coatings, textiles, food, cosmetics, and medicine, and is certain to have a major influence on future society. There are many potential applications, but it is important to carry out rigorous safety and toxicology tests on nanomaterials before they are introduced on a large scale. For example, what effect do they have on human health if they are inhaled, swallowed, or absorbed through the skin? Could they irritate the lungs and cause damage similar to the effects of breathing in fine dust? What effect might nanoparticles have on the human immune system? If large quantities of nanoparticles enter the environment, how would that affect insects, birds, fish, and animals? Finally, how might nanotechnology be misused by criminals, terrorists, and unscrupulous institutions?

These questions have already been raised, and so properly designed tests need to be carried out to assess if there are any risks. Unfortunately, many of the toxicological studies carried out so far have been flawed because of the excessive quantity of material tested. Proper toxicology testing should establish whether a material is safe under realistic conditions and concentrations. For example, table salt can be shown to be toxic in high doses, but nobody would seriously consider removing it from supermarket shelves. To that end, there has been some discussion about introducing a regulatory system that would oversee nanotechnology. The EU has already issued guidelines (in 2011) on how toxicology tests should be carried out on nanoparticles.

Further reading

Organic Chemistry, 2nd edition, 2012, by Jonathan Clayden, Nick Greeves, and Stuart Warren, Oxford University Press.

Organic Chemistry, 8th edition, 2011, by John E. McMurry, Brooks/Cole.

Foundations of Organic Chemistry (Oxford Chemistry Primers), 1997, by Michael Hornby, Oxford University Press.

Beginning Organic Chemistry (Workbooks in Chemistry), 1997, by Graham L. Patrick, Oxford University Press.

Organic Chemistry I for Dummies, 2016, by Arthur Winter, John Wiley and Sons.

Organic Chemistry II for Dummies, 2010, by John T. Moore, John Wiley and Sons.

BIOS Instant notes in Organic Chemistry, 2003, by Graham Patrick, Taylor and Francis.

An Introduction to Drug Synthesis, 2015, by Graham Patrick, Oxford University Press.

Index

Index

SCIENTIFIC REVOLUTION
A Very Short Introduction
Lawrence M. Principe

In this *Very Short Introduction* Lawrence M. Principe explores the exciting developments in the sciences of the stars (astronomy, astrology, and cosmology), the sciences of earth (geography, geology, hydraulics, pneumatics), the sciences of matter and motion (alchemy, chemistry, kinematics, physics), the sciences of life (medicine, anatomy, biology, zoology), and much more. The story is told from the perspective of the historical characters themselves, emphasizing their background, context, reasoning, and motivations, and dispelling well-worn myths about the history of science.

THE HISTORY OF MEDICINE
A Very Short Introduction
William Bynum

Against the backdrop of unprecedented concern for the future of health care, this Very Short Introduction surveys the history of medicine from classical times to the present. Focussing on the key turning points in the history of Western medicine, such as the advent of hospitals and the rise of experimental medicine, Bill Bynum offers insights into medicine's past, while at the same time engaging with contemporary issues, discoveries, and controversies.

www.oup.com/vsi

NUCLEAR POWER
A Very Short Introduction
Maxwell Irvine

The term 'nuclear power' causes anxiety in many people and there is confusion concerning the nature and extent of the associated risks. Here, Maxwell Irvine presents a concise introduction to the development of nuclear physics leading up to the emergence of the nuclear power industry. He discusses the nature of nuclear energy and deals with various aspects of public concern, considering the risks of nuclear safety, the cost of its development, and waste disposal. Dispelling some of the widespread confusion about nuclear energy, Irvine considers the relevance of nuclear power, the potential of nuclear fusion, and encourages informed debate about its potential.

www.oup.com/vsi